Effects of Welding on Health II

An up-dated (January 1978-May 1979) literature survey and evaluation of the data recorded since the publication of the first report, to understand and improve the occupational health of welding personnel.

Research performed at the Franklin Research Center under contract with the American Welding Society and supported by industry contributions.

Prepared for:
SAFETY AND HEALTH COMMITTEE
AMERICAN WELDING SOCIETY
2501 NW 7th Street, Miami, Florida 33125

Library of Congress Number: 79-5267
International Standard Book Number: 0-87171-212-

American Welding Society, 2501 N.W. 7th Street, Miami, FL 3312

Note: By publication of this document the American Welding Society does not insure anyone utilizing the documen against liability arising from the use of such document. A publication of a document by the American Welding Societ does not carry with it any right to make, use, or sell any patented items. Each prospective user should make a independent investigation.

Printed in the United States of Americ

Contents

Personnel

Authors of the report by Franklin Research Center were:

Samir Zakhari and Roy S. Anderson

AWS Research Committee

J.F. Hinrichs, Chairman	A.O. Smith Corporation
A. Lesnewich, Vice-Chairman	Airco Welding Products
M.E. Kennebeck, Jr., Secretary	American Welding Society
K.L. Brown	Lincoln Electric Company
R.E. Kinser	Caterpillar Tractor Company
P.C. Krueger	Airco Welding Products
P.W. Ramsey	A.O. Smith Corporation
A.N. Ward	Caterpillar Tractor Company

Acknowledgments

The American Welding Society gratefully acknowledges the time and effort expended by the members of the Research Committee and the Research Finance Committee and the financial support of the program by industry contributions.

Supporting Organizations

Alloy Rods
  Division of Allegheny Ludlum Steel Company
Westinghouse Electric Company
Airco, Inc.
Miller Electric Company
The Heil Company
United States Steel Corporation
Union Carbide Corporation
Babcock and Wilcox Company
General Electric Company
E.I. DuPont de Nemours & Company
Teledyne McKay, Inc.
J.I. Case Company
Lincoln Electric Company
Huntington Alloys, Inc.
Stoody Company
AWS Detroit Section
Duke Power Company
Sandvik, Inc.

Many other organizations have made contributions to support the ongoing program from April 1979 to the present.

Preface

This literature review has been prepared for the Safety and Health Committee of the American Welding Society to provide an assessment of current knowledge of the effects of welding on health, as well as to aid in the formulation of a research program in this area, as part of an ongoing program sponsored by the Committee. Previous work has included studies of fumes and gases, radiation, and noise generated during various forms of arc welding. Conclusions based on this review and recommendations for further research are presented in the introductory portions of the report. Chapter 1 summarizes the occupational exposures. Chapters 2 and 3 contain information related to the effects of exposure to byproducts of welding operations on humans and laboratory animals. Chapter 4 covers *in vitro* studies.

Referenced materials are available from the Franklin Institute.

Introduction

The health of welders has always been a concern of the American Welding Society (AWS). Much of the earlier literature has been reviewed by The Franklin Research Center (FRC); findings were presented in October 1978 to the Safety and Health Committee of AWS in the report *Effects of Welding on Health* (Ref. 1).

Since then, much scientific study has been reported, and it was found necessary to update the earlier report. As the present work focuses mainly on health effects of welding, newly published methods of sampling and analyses of welding fume have not been emphasized.

Like workers in the metals industry, welders are generally exposed to a variety of occupational risks such as cuts, bruises, and sometimes more severe injuries due to the handling of massive metal objects. Furthermore, due to the nature of their work, welders are exposed to other occupational hazards such as heat, radiation, and the inhalation of fumes that might cause acute or chronic health effects. It is the purpose of this review to evaluate and present an update to include information on the effects of welding on health published from the start of 1978 through May 1979.

The reader is cautioned that the papers reviewed were examined only for manual consistency. No independent checks of the experiments were performed. This report must be read in conjunction with *Effects of Welding on Health* by Villaume et al.

Executive Summary

Although welders are exposed to various potentially hazardous substances, overt clinical disease is an infrequent phenomenon among adult welders. This is due to continually improved industrial hygiene techniques, research findings, and the fact that repeated exposure to a given compound in small concentrations could stimulate various defense mechanisms in the human body to effectively deal with the "invader." Tolerance (that is, reduced sensitivity of the body to the toxic effects of chemicals) and the development of various immune responses are factors that contribute to the lack of overt toxicity syndromes in certain cases. On the other hand, life-style factors, such as smoking or drinking habits, use of drugs, and personal hygiene, could not only change the clinical picture of metal exposure, but also give rise to totally misleading findings. It follows, therefore, that we are faced with two situations that might lead to erroneous conclusions: first, subclinical cases that are considered normal on medical examinations; second, clinical cases that are inaccurately blamed on welding exposure. However, these two groups have in their biological systems a variable amount of "extrinsic substances." It is clear, therefore, that biological monitoring for the presence of various chemicals in biological fluids or tissues, or both, constitutes one accurate measure by which health personnel can estimate the level of exposure to a given toxic substance, particularly the complex mixture known to be involved in welding processes, especially when used in conjunction with careful environmental monitoring techniques.

Regular testing of the urine, blood, hair, or enzymatic patterns of welders could be used to alert the industrial hygienist to the possibility of overexposure to a given compound. There are two limitations to the use of biochemical testing:

(1) The only criterion in such analysis is the deviation from what would be expected to be "normal." Certain biological tests, especially those that involve enzymatic systems, have a wide range of "normal" or control values.

(2) The toxicokinetics (that is, the absorption, distribution, and excretion) of different metals do not obey a fixed pattern of behavior; thus, while some metals are cumulative, others are not. With the development of modern technology, it would be very gratifying to develop sensitive and specific diagnostic tests to detect any ailment in welders long before any "irreversible" damage might occur.

Before recommending specific biochemical tests for determining the extent of welder exposure to fumes, it was found necessary to review the merits of the analysis of various heavy metals in biological fluids.

Analyses for Metallic Compounds

It is becoming increasingly clear that analyses of blood, urine, expired air, and other specimens of biological origin, though not always reliable, have value in conjunction with environmental monitoring to determine the degree of exposure to environmental substances. The main justification for this joint approach lies in the fact that direct analysis of environmental air, though providing an accurate assessment of the amount of potentially toxic substances in ambient air, fails to precisely reflect the degree to which these materials are actually absorbed by different biological systems. Since the degree of toxicity is directly related to the amount of toxic substance delivered to the target organ, it follows that detection of these substances in biospecimens potentially provides a better measurement of exposure and, hence, the degree of toxicity. It, is, therefore, suggested that the joint determination of the amount of metals or gases, or both, and their metabolities in biospecimens from welders (such as blood, urine, hair, nails, saliva), plus environmental monitoring, could be used to measure the extent of exposure, and better estimate the total risk.

Urine Analysis for Metallic Compounds

Generally speaking, urine analysis constitutes a practical method for monitoring the extent of exposure. Its value in the diagnosis of exposure to certain metals is dubious. Urine analysis of workers exposed to various metals showed that significant excretion in excess has been observed for such elements as nickel (Refs. 2 and 3); chromium (Refs. 4 and 5); cadmium (Refs. 6, 7, and 8); lead (Refs. 9 and 10); arsenic (Ref. 11); fluorides (Refs. 12, 13, and 14); and selenium (Ref. 15). Some observations in recent publications are summarized in Table 1.

Blood Analysis for Metallic Compounds

In 1977, Ulfvarson and Wold (Ref. 16) estimated the concentration of 17 trace elements (lead, strontium, rubidium, beryllium, gallium, zinc, copper, cobalt, iron, manganese, chromium, calcium, potassium, sulfur, phosphorus, silicon, and magnesium) in whole blood samples

Table 1
Observations from recently published urine analysis studies

Metal	Occupation	Observations	Reference
Nickel	Welders	1. Correlation between exposure and excretion was demonstrated for levels 1 to 5 mg/m^3. 2. No correlation was found in welders exposed to less than 1 mg/m^3. 3. No increase in urinary Ni excretion was found in welders exposed to 0.1 mg/m^3.	3. Norseth and Gundersen 1978
		1. Moderate increase in urinary Ni excretion was observed in arc welders. 2. Measurement of urinary Ni is more sensitive and practical than that in serum for determining the extent of exposure.	185. Bernacki and Parsons 1978
Chromium	Welders	1. Urinary Cr level is a good indicator of short term exposure.	4. Tola et al. 1977
		2. A high degree of correlation was found between inhaled Cr and its concentration in urine.	5. Gylseth et al. 1977
Lead	Welders	1. Exposure to Pb for short times increased its level in urine; prolonged exposure resulted in a less prominent increase. 2. Urinary ALA levels were increased after exposure to lead.	9. Cramer et al. 1973

of 71 persons utilizing various welding methods. They could not find a correlation between the level of exposure and blood levels. However, other authors found that the serum level of various metals is substantially increased in workers exposed to metals such as zinc (Ref. 17), copper, chromium, and nickel (Refs. 18 and 19), lead (Ref. 20), and cadmium (Ref. 6). Various observations are shown in Table 2.

The "normal" amount of various metals in urine and blood as well as methods of their determination are mentioned elsewhere (Refs. 22 through 26).

Perspiration Analysis for Nickel

Sunderman (Ref. 27), in a review of the toxicity of nickel, maintained that the concentration of nickel in perspiration was approximately 20 times greater than that in urine. If this is the case, determination of nickel in perspiration may constitute a useful method for the estimation of exposure.

Biochemical Tests for Metallic Exposure

Biochemical tests show changes in the function of various organs due to excessive exposure to a given compound. Nonetheless, due to the wide range of "normal" values, slight fluctuations in these values due to early toxicity cannot be easily detected.

Recently, some specific and highly sensitive tests have been developed to detect dysfunction of the target organs in early stages of poisoning. Such tests are exemplified by the determination of serum ornithine carbamyl transferase (OCT) levels and greyscale ultrasonography as tests for the detection of early liver damage. It is well known that antimony, beryllium, cadmium, cobalt, copper, and molybdenum bring about liver damage in humans. The effects of metallic elements usually present in a welding atmosphere on serum protein composition in humans are summarized in Table 3. Tables 4 and 5 summarize findings in experimental animals (modified from DeBruin, Ref. 28).

Table 2
Observations from recently published blood analysis studies

Metal	Occupation	Observations	Reference
Chromium	Autoworkers	Cr blood levels were significantly high.	18. Clausen and Rastogi 1977
Lead	Autoworkers	1. Increased blood Pb levels was observed in 59 percent of workers. 2. Significant correlation between blood Pb levels and ALA-D activity was found.	19. Clausen and Rastogi 1977
Manganese	Workers in Mn alloy plant	Mn concentration in blood did not increase with the increase in exposure time.	21. Tsalev et al. 1977
Nickel	Autoworkers	Ni blood levels were significantly high.	18. Clausen and Rastogi 1977
Zinc	Various occupations with high oxide exposure	1. Serum Zn concentrations are high and could be used to diagnose occupational exposure. 2. Serum Zn level is high in patients with zinc fever, especially on the first day.	17. Jaremin 1977

Table 3
Serum protein composition in human exposure

Compound	Exposure condition	Observations	Interpretation
Lead (inorg.)	Occupational workers, prolonged exposure	General hypoalbuminemia with a corresponding rise of alpha and beta globulins in chronic plumbism; behavior of gamma globulin is not uniform; values either above or below normal were encountered. Slight protein shifts may occur in the absence of distinct signs of toxicity; changes are more pronounced with increasing severity of clinical signs and with length of contact.	Liver insufficiency, shown by functional tests; possible RES involvement.
Cadmium	Occup. exp.	Decline of albumin; rise of various globulins.	Hemolytic action?
Manganese	Occup. exp., chronic	Patients show a fall in albumin, and a rise in alpha, alpha$_2$ and gamma globulin; hyproproteinemia.	Probable liver involvement.
Beryllium	Chronic poison	Fall in A/G in chronic berylliosis; globulin rise; the course and stage of disease; hypergammaglobulinemia.	Etiology not defined; fall in blood SH.
Selenium	Se factory	Decline of total protein content.	
Copper	Cu mining plants	Dysproteinemia, most marked in subjects with signs of poisoning.	Hematological disorders (rise in blood Cu, Fe).
Molybdenum	Chronic exp. (CuMo plant)	Fall in A/G ratio; rise of globulins, particularly the alpha fraction.	Liver dysfunction.
Nickelcarbonyl, Ni $(CO)_4$		Dysproteinemia; marked rise of alpha globulin.	Lung edema; hematological changes.
Carbon monoxide, CO	Occup. exp., coke furnace workers (a.s.o.)	Marked shift from albumin to globulin, notably beta globulin in heavy prolonged exposure (500 ppm and more).	Etiology not well defined.

Recommended Tests

1. Analysis of urine for:

(a) *Chromium.* Occupational exposure to chromium is reflected by an immediate increase in its concentration in urine. However, both blood and urinary content of chromium remain elevated for several years after the cessation of exposure.

(b) *Fluorides.* The concentration of fluorides in human urine is closely correlated to the amount that obtains access to the body either by the oral route or by inhalation. A characteristic feature of fluoride exposure is the short onset time for its appearance in urine. Urine analysis is, therefore, considered to be the best means of estimating fluoride exposure.

2. Analysis of blood for:

(a) *Copper.* Determination of copper in blood is a more reliable method for assessing the degree of exposure than is estimating its level in urine.

(b) *Lead.* Serum and urine levels of lead are reliable measures in the diagnosis of early lead exposure. Prolonged exposure to lead, however, produced lower urinary lead levels than immediate exposure (Ref. 8). Nonetheless, there is a high correlation between lead levels in ambient air and that in body fluids. This correlation permits the determination of the level of exposure and allows predictions of the total lead body burden. A more accurate diagnostic test for the severity of lead exposure is the estimation of the activity of ALAD (lead causes a decrease in ALAD level that is not related to smoking).

3. Analysis of hair for:

(a) *Lead.*

(b) *Cadmium.* It seems that there is no evidence of a quantitative correlation between the urinary cadmium level and either the degree or the duration of exposure. The amount of cadmium in urine increases significantly during continuous exposure and persists for several years

Table 4
Effect of elemental compounds on serum enzymes

Compound(s)	Conditions of treatment (species)	Enzyme(s) tested	Effect
Lead (Pb) salts	Multiple treatment, rabbit	GOT, GPT	Inc.
	Acute treatment	CPK, hexokinase	Inc.
	Chronic pois. sheep	GPT	Inc.
	Repeated i.p. injection, rat	LDH	Inc.
	Oral dosage, guinea pig	LDH, GOT-1 & 2, G1-DH	Inc.
	Acute i.v. treatment	Alk. Ph.	Dec.
	Chronic adm., sheep, rat	Pseudo Ch.E	Dec.
Beryllium (Be) salts	Chronic dosage	Als. Ph.	Inc.
	Chronic treatment, rabbit	Transaminase, MDH, ICDH, rat	Inc.
	Lethal poisoning intra-tracheal injection	Alk. Ph.	Inc.
Manganese (Mn) compounds	Repeated i.v. dosage	Transaminase LDH	Inc.
		Adenosine deaminase	Inc.
		Pseudo Ch.E	Dec.
Cobalt (Co) salts	Heart damaging doses, rat	Transaminase, ALD, CPK, LDH	Inc.
Cadmium (Cd) salts	Multiple high dosage	Transaminase	Inc.
		Pseudo Ch.E	Inc.
	Repeated doses	GOT, GPT, LDH	Inc.
	Chronic pois.	GOT, G1-DH	Inc.
	Subacute treatment	Alk. Ph.	Inc.
	Chronic, s.c. doses	Amylase	Inc.
Copper (Cu) salts	Repeated adm., cattle	Transaminase LDH	Inc.
Zinc (Zn) salts	Various treatments	LDH	Dec.
Molybdenum (Mo) salts	Acute intake	Pseudo Ch.E	Dec.
		Alk. Ph.	Dec.
Nickel (Ni) chloride	Chronic, i.v. intoxication, rabbit	GOT, GPT, LDH, ALD	Inc.
Selenium (Se) salts	Varying doses	Pseudo Ch.E	Dec.
Inorganic fluoride (F)	Fluorotic dosage, cattle, chicken, rabbit	Alk. Ph.	Inc.
Carbon monoxide (CO)		ALD, CKP, GOT, LDH, MDH & SHD Pseudo Ch.E	Inc.

Table 5
Serum lipoprotein changes in animal poisoning

Compound	Animal species	Treatment, condition	Result
Lead acetate	Rabbit Rat	Long term i.e. administration Intraperitoneal application	Increase of beta lipoprotein, with associated hypercholesterolemia
Zinc salt	Rabbit	Acute doses	Shift from alpha to beta fraction; rise of beta/alpha ratio
Cadmium salt		Chronic doses	Increase of beta/alpha ratio
Carbon monoxide	Rabbit	Chronic exposure (2500 h, 100 ppm)	Rise in beta/alpha quotient
Silica	Guinea pig	Intraperitoneal	Rise in beta lipoprotein; hyperlipemia
Quartz powder	Dog, rabbit	Intrabronchially, intravenously	Rise in beta lipoprotein fraction

after the cessation of exposure. Therefore, little or no information can be obtained from blood or urinary levels of cadmium; however, this can be used as a rough estimate of the amount of cadmium stored in the body. Hair samples are, therefore, recommended to estimate the exposure.

(c) *Nickel.*

4. Fecal analysis for manganese.

The amount of manganese excreted in urine represents but a small fraction of the total amount in the body. It follows that the value of the determination of manganese in urine as an index of the exposure level is questionable. The blood level in workers exposed to manganese could be as high as four times the normal value. Determination of blood levels of manganese is of greater significance than that of urine. Manganese exposure is claimed to be accurately determined by estimating its level in the feces (Ref. 29). This is dependent on the fact that manganese is preferentially excreted via bile.

Technical Summary

According to a recent population survey in the U.S. (unpublished report, Bureau of Labor Statistics 1979), there are an estimated 713,000 welders and flame cutters who are actively employed. Welders are exposed, in varying degrees, to occupational hazards such as cuts, bruises, and more severe injuries involved in the handling of metal objects. Furthermore, because of the nature of their work, welders are exposed to heat, radiation, noise, and welding fumes that may cause acute or chronic health effects.

It is the objective of this document to update an earlier report titled *Effects of Welding on Health,* to evaluate the present state of knowledge of the effects of welding on health, and to provide recommendations to the American Welding Society for future research. To accomplish these three objectives, scientific studies published between the start of 1978 and May 1979 were evaluated. The information was evaluated and arranged in four chapters: The Exposure, Effect of Welding on Human Health, Toxicologic Investigations in Animals, and In Vitro Studies.

The Exposure

In this chapter, studies performed on welding fumes, gases, radiation, and noise are summarized.

Fumes

The objective of engineers and industrial hygienists is to reduce the concentration of fumes in the welder's environment. This goal can be achieved by: (a) use of welding methods that produce the least volume of fumes, and (b) adjustment of the rate at which fumes are removed, i.e., ventilation requirements. The use of inert gas welding was found to reduce the diffusion of metal vapor and, hence, fume generation, as shown by Kobayashi et al. (Refs. 30 and 31). From an industrial hygiene viewpoint, the publications of Stern (Refs. 32 through 35) are of interest. In a series of articles, Stern suggested a classification of welding electrodes according to what he called the "electrode profile." Each type of electrode, according to Stern, should carry a label showing its productivity in kg/hr as compared to the air flow (in m^3/sec) required to bring the generated fumes to the recommended TLV's. A formula for calculating the ventilation requirements was developed by Magnusson (Ref. 36).

Gases

The sources of gases that are apt to be produced during welding are discussed. Gases cited are: nitrogen oxides, ozone, carbon monoxide, and, under certain unique conditions, phosgene or phosphine.

Radiation

Welders are exposed to three types of radiation: ultraviolet, visible, and infrared. Hinrichs (Ref. 37) maintained that, as in the case of fumes, the extent of radiation differs from one welding process to the other. Highest emissions were found in gas metal arc welding and the lowest in plasma arc welding.

Visible light produced during welding was found to be predominantly in the short wave (420 to 430 nm) spectrum. Zaborski (Ref. 38) attributed these emissions to atomic and molecular radiations from vapors and gases in the arc. Ozone is generated by ultraviolet emission at some distance from the arc. Exposure to ultraviolet light could cause conjunctivitis, or "welder's flash." Shade shield recommendations by NIOSH (ANSI Standard Z87.1-78) should be followed. It is important to recognize that certain chemicals, such as the furocoumarines and psoralens, present in various foods and drugs have the capacity to increase photosensitization to UV radiations. Ray burns of the skin should be interpreted not only in terms of light intensity but also in terms of possible photosensitization.

The most dangerous effect of infrared radiation is the development of cataracts, which was found to be caused by wave lengths below 1400 nm.

Noise

Excessive exposure to noise could be a substantial health hazard to welders. Noise is associated with other shop operations as well as with welding.

Effects of Welding on Human Health

In this chapter, the effects of welding fumes on human physiological systems are discussed. The data about the possible carcinogenicity of welding fumes, human fatalities, and recent epidemiological studies are summarized and evaluated.

Background

It has been shown by Tierney (Ref. 39) that health hazards associated with welding fumes constitute only a small fraction of the total occupational risk of welding. Hazards from mechanical or accidental events that, fortunately, could be easily eliminated constitute the major bulk of risk. The scope of this report is limited to the effects of welding fumes on health.

Toxicity to Various Organs

Respiratory System

Factors that govern the toxicity of gases and particulates in welding fumes and various defense mechanisms that are available to the body to expel these undesirable substances were reviewed. The well known inhibition of the body's defense system by tobacco smoke and the synergistic effects of smoking acting in concert with other materials can lead to lung dysfunctions of varied types.

Reported cases of pneumoconiosis from particulates such as iron, aluminum, copper, and beryllium were discussed. The introduction of scanning electron microscopy enabled Guidotti and DeNee (Ref. 40) to study metal particles that were not otherwise visible. Using the same technique, Eskilsson et al. (Ref. 41) found that the most biologically active fumes contained particles rich in chromium often associated with potassium.

Whether or not aluminum and iron particles induce interstitial fibrosis in the lungs remains an unresolved question. Thus, while Patel et al. (Ref. 42) claimed that iron oxide is fibrogenic, Stettler et al. (Ref. 43) maintained that it is not.

Agents that produce pulmonary irritation, such as nitrogen oxides, ozone, and cadmium vapor, were discussed. Only one case of acute pulmonary edema was attributed to nitrogen oxides and ozone. The case involved the use of oxyacetylene and tungsten arc welding in a confined space. The patient was treated with ampicillin and prednisone, and complete recovery ensued in eleven days. Cadmium fumes, on the other hand, induced pulmonary edema that ended fatally in one case (Ref 186).

Lung function tests were used to study the effect of welding fumes on the lung. Since lung function tests are insensitive measures of lung dysfunction, test results may appear normal in the presence of other manifestations of disease. Despite this problem, the lung function tests are especially useful in the early detection of the pneumoconiosis.

Eye and Vision

The effects of fumes and radiation on welder's eyes were discussed. It appears that ultraviolet radiation is the major source of complaints. Ross (Ref. 44) examined 926 welders over a six-year period. He could find no long-term effects on either near or far vision. He reported only one case of cataracts that was unrelated to infrared exposure.

Ears and Hearing

An average hearing loss of 34 dB in the speech frequency range was reported (Ref. 44). Adequate protection against noise is necessary.

Skin

Burns from hot metal and ultraviolet radiation are quite common among welders. In view of the material's meltability, nylon clothing does not provide adequate skin protection.

Gastrointestinal Tract

Welders using low hydrogen electrodes complained of loss of appetite and stomach ache (Ref. 45).

Cardiovascular System

Although dynamic arterial hypertension was reported by Gola and Galazka (Ref. 46), a detailed study is needed to confirm the effect of fumes on the cardiovascular system.

Blood

Nickel, cobalt, chromium, iron, and molybdenum induced a hemolytic effect on human erythrocytes *in vitro* (Ref. 47). Only one case of granulocytopenia was reported in a welder who was employed for 20 years (Ref. 46). However, since several drugs and chemicals are able to induce such an effect, the association of granulocytopenia to welding fumes is questionable. Medicaments that might have been used by that welder (a 39-year old) were not discussed.

Central Nervous System

Manganese, a component of welding fumes, seems to cause certain visual motor disturbances and reduction in short-term memory acuity. Although cases of definite neurological manifestations were reported in workers in a battery factory (not welders), no studies indicating that this is a problem in welders have been found recently. Past reports show that chronic manganism is seen only rarely in welders.

Liver

Exposure to various metals and their oxides in welding fumes resulted in certain biochemical changes in the liver that are discussed below under the heading Biochemical Changes.

Urinary System

Cases of urinary system involvement in welders were not published in 1978-79. However, exposure to cadmium and lead fumes produced damage to the proximal tubules of the kidney.

Allergic Reactions (Metal Fume Fever)

Generally, more febrile episodes were encountered in welders than in nonwelders. This is due to exposure to welding fumes, especially those that are rich in zinc oxide. Removal of the patients from the polluted environment and supportive measures constitute the main line of treatment.

Biochemical Changes

The most consistent biochemical change observed was the decrease in the activity of δ-aminolevulinic acid dehydratase due to the exposure to fumes rich in lead. This is a reliable test because the result is not affected by smoking. Although serum and urine levels of lead were also increased, lead determination in these fluids is a reliable measure in the diagnosis of early lead exposure. Prolonged exposure produced lower urinary lead levels than immediate exposure.

For the 1978-1979 period, no published research on the effect of welding fume on the musculoskeletal, reproductive and endocrine systems, or on teeth and the oral cavity was found.

Carcinogenicity of the Fumes

The reported induction of cancer due to the occupation of welding will be mentioned briefly. To guard against hasty conclusions, each case will be followed by a short comment.

Skin Cancer

The case reported by Raynor et al. (Ref. 48) involved a 59-year old welder who suffered a traumatic injury of the left thumbnail in which a "black welding flux" was introduced under the nail. An epitheliod cell malignant tumor was observed 13 years later in the dorsum of the distal phalanx of the involved thumb. Neither a detailed history of the patient's occupation before and after the accident nor the chemical composition of the "black flux" were given.

Examination of this case throws a great deal of uncertainty on the causative agent. In the absence of a chemical analysis of the "black flux," it would be very difficult to judge whether or not it was the agent that caused the cancer. It is well known that certain substances such as arsenic, cutting oil, and tars are capable of inducing skin cancer. Exposure of the traumatized tissues to any of the cancer-causing agents, either during work or at home, could have caused the development of cancer. An analysis of the previous occupational exposure of this patient and his activities during the 13 years would be necessary to reach a firm conclusion.

Brain Tumors

In this case, the patient was a 52-year old welder who had spent 22 years as an electric welder; prior to that, he had been a coal miner. Before admission to the hospital he spent two weeks welding zinc coated pipes. He then developed symptoms typical of metal fume fever. Analysis of serum-metal levels showed a fivefold elevation in zinc. The autopsy showed bronchial problems as well as "black lung" indications, tumors in the parietal lobe, and in the left hemisphere (Ref. 17).

Zinc is normally present in the brain, and it is unlikely that a two-week exposure to zinc fumes could have been the sole cause of the brain tumors. It is technically justifiable to assume that excessive zinc absorption might have contributed to the unveiling of latent clinical symptoms of preexisting brain tumors of unknown etiology.

Cancer in the Lung and Nasal Sinuses

Reported cases of lung and nasal sinus cancer were due to exposure to nickel in a nickel refinery (Ref. 49), but none have been reported in welders. Although nickel is present in welding fumes, there is no way of extrapolating the above results to welders because of the unknown interactions of nickel, iron, and other metals, gases, etc., that occur during welding. The evidence appears to show that typical nickel nasal erosions do not occur in welders.

Cancer of the Bladder

Perhaps the epidemiological studies performed by Milham (Ref. 50) offer the best framework for a study of cancer risk in welders. In his study, Milham analyzed the mortality patterns associated with exposure to metals in Washington State in the period 1950 to 1971. He concluded that welders and flame cutters develop more lung cancer and lung related diseases than expected. Welders also showed a high, though statistically nonsignificant, proportionate mortality ratio from cancer of the urinary bladder. Although this study is indicative, it is not definitive, and complete occupational exposure history of these people should be studied before reaching firm conclusions. Continuous irritation as from bladder stones could theoretically initiate tumorigenesis. Exposure to known cancer-producing substances, such as benzidine or beta-naphthylamine, should also be taken into account.

Toxicokinesis of the Fume

The fate of fluorides in welders' bodies was studied by Pantucek (Ref. 51). He found a close correlation between the concentrations of fluorides in air and that in urine of welders.

Epidemiologic Studies

Epidemiologic studies conducted by Buncher et al. in 1977 (Ref. 52) revealed that skin injuries and the occurrence of cough in the morning in welders are the most significant findings. The epidemiologic study of Milham (Ref. 50) was mentioned earlier.

Human Fatalities

Four fatalities were attributed to welding (Refs. 17, 44, 48, and 53). The duration of exposure ranged from 5 to 22 years and the causes of death were lung related.

Toxicologic Investigations in Animals

Only four papers appeared in 1978-1979 that discuss the effect of welding fumes on experimental animals. The bulk of the literature shows that only single metallic components were used for the study. This is unfortunate, since results from such experiments are in no way applicable to the welding fumes. Six reasons were offered at the start of this summary to explain why such experiments do not add to the understanding of the fume toxicity. However, these experiments were included in an attempt to find possible correlations between pathological lesions from metals and their level in biological fluids.

In Vitro Studies

Stern (Refs. 33 and 34) and Maxild et al. (Ref. 54), using Salmonella to test for the mutagenicity of welding fumes, found that those fumes that evolved were in some instances mutagenic and in others not mutagenic. See the initial report, *Effects of Welding on Health*, p. 55, for a complete discussion of Maxild's results. Stern's work and Maxild's are essentially similar.

Recommendations

Reports of human exposure published in the last two years show that a large amount of information on the health effects of welding has been gathered. The information is incomplete, some is questionable, and areas remain where more study and information are needed in order to develop methods for protecting welders. The following recommendations are submitted for consideration.

Epidemiologic Studies

Epidemiologic studies have provided a great deal of knowledge about the toxicity of various substances such as lead, arsenic, and asbestos in the workplace. Since epidemiologic data are derived from observations on human populations, such studies are far more significant than animal studies. There are numerous variables that, if left unaccounted for in the analysis of any association between a given substance and the resulting health effect, may lead to either a false positive or a misestimation of the magnitude of the risk. For the occupation of welding, these variables include not only the welding fumes, which vary dramatically in their composition from minute to minute, but also exposure to other agents in the occupational setting, the life-style, and habits such as cigarette smoking, street drug usage, and dietary patterns. It is recommended that future epidemiologic studies carefully consider the following points:

(1) Although it is known that the most prevalent (obvious) health effects in the welding population are those that involve the eye and lung, any study must take into account other possible effects, positive or negative, on various physiological systems of welders. This would include considerations of such diseases as dermatoses, cataracts, coronary heart diseases, dental involvement, hearing loss, and cancer. For example, an unprotected welder could suffer a loss of hearing with a subsequent social handicap and a reduction in his quality of life, although precautions may have been taken to protect other organs such as eyes and lungs.

(2) A query to NIOSH's National Occupational Hazard Survey Data Base yielded a list of about 1,000 agents to which welders and flame cutters (code 680) were exposed during the survey. Analyses of epidemiologic data should take into account other agents (that are known to be carcinogenic in man such as asbestos) to which welders were exposed in addition to the welding exposure. Cancer could be caused by these agents and erroneously attributed to welding.

(3) Obviously, smoking and other life-style factors of the welders studied must be taken into consideration.

(4) Reliable records should be kept of the type of metal(s) that is being welded, the method used (including the flux or shielding gas), the welding rods or electrodes (including coating or contaminants), as well as other agents such as degreasing solvents. The composition of fume should be analyzed, and industrial hygiene measurements of the workplace atmosphere should be performed periodically and recorded. Ideally, a detailed history similar to the 24-hour overall history used in nutritional studies should be applied to an assessment of past occupational exposure.

(5) Although the concentration of pollutants in air can be accurately measured, this may not reflect the actual dosage received or the body burden. Periodic testing of blood, urine, hair, and feces for levels of metals such as chromium, nickel, cadmium, etc., should be performed; results of these tests should be considered during analysis of the epidemiological findings. A health effect could be associated with the presence of metals in the body, when enough data has been collected to show what normal variations could be expected and which were excursions that indicate a start of potential damage.

Experimental Studies

Metals such as arsenic, cadmium, cobalt, copper, lead, manganese, molybdenum, nickel, and zinc are teratogenic in experimental animals, but no mention in the literature of the teratogenic potential of welding fumes that contain these metals was found. To take appropriate measures for protecting female workers who are exposed to the fumes during childbearing age, it will be necessary to study any potential teratogenic effect of welding fumes in experimental animals.

Reproductive effects have not been sought in human epidemiological studies or in experiments with animal models. Such a study might be initiated to learn if fertility or other changes are evident among welders.

Interferon (a substance that plays an important role in cellular defense mechanisms such as phagocytosis) production in humans is known to be depressed by certain dust particles. Interferon is one of the defense mechanisms of the lung that may be breached while inhaling welding fumes, with a possible higher occurrence of lung diseases. It therefore would be of great interest, at least from an academic viewpoint, to study the interrelationship between the presence of interferon and welding fumes. This study may lead to a better understanding of the role of interferon in protecting the lungs from possible damage and in clearing the lungs of both metal and organic particles.

Safety Measures

Work practices in the U.S. have gone far to ensure the protection of welders from the hazards of their work. Good industrial hygiene practices, the continued improvement of personal protective devices, the health and safety standards, and the studies established and performed by the American Welding Society and by other technical societies in the U.K., Sweden, and Japan have played a large part in reducing the risk due to welding in the industrial environment.

Chapter 1

The Exposure

Various welding methods, as well as the composition of the fumes generated during welding, are described in detail in an earlier report (Ref. 1). About 15 months have elapsed since the submission of that report, during which time, according to our best knowledge, there were no breakthroughs in the introduction of new welding methods; nor have there been any extraordinary findings concerning the composition of fumes. However, it is appropriate to examine some basic investigations on various mechanisms that govern the formation of welding fumes in order to formulate the appropriate control measures to combat their effects on the health of welders.

Fumes

Direct observation, using a high-speed camera, of the fume generation stage of arc welding was reported by Kobayashi et al. (Ref. 31). The investigators found that the amount of fume generated was dependent upon the intensity of the welding current, reaching a maximum at the current maxima. This is due to the fact that an increase in electric current will increase the temperature at the vaporizing surface, which, in turn, increases the amount of high temperature vapors that form the fumes. The blowing velocity of the high temperature vapor reaches its highest rate when the current is at its maximum. It was also shown that most fume is generated from the lower part of the arc column. This effect was ascribed to the highest temperature of the vaporizing surface being in that area. An excellent review of the nature and origin of particles in condensation fumes was published by Buckle (Ref. 55).

The correlation between the total amount of fume produced and the welding current used was studied by Stern (Ref. 32). In his report to the Danish Welding Institute, he found that there is a linear relation between the total fume produced and the voltage at constant current. On the other hand, at constant voltage, the amount of fume produced per electrode is independent of current used. This finding is very important because it follows that the ventilation requirements are dependent upon the power consumed in the welding process. Based on these findings, Stern suggested a classification of welding electrodes according to what he called the "electrode profile." In this system of classification, the ratio between the productivity of a given electrode, measured in kg/hr, and the ventilation requirement, in terms of m^3/sec of air, is compared to the power requirements. Similar results were found by Kayano et al. (Ref. 56) in aluminum arc welding.

In another investigation, Senn (Ref. 57) studied the volume of beryllium and copper fumes generated during welding BeCu plate. Senn found the welding fume generation rate (μg/welding minute) to be as follows:

Type of BeCu plate	Fume generation rate (μg/m) (micrograms per welding minute) Be	Cu
1.8% Be 0.95 cm (3/8 in.) thick, not age-hardened	195	2,330
0.5% Be 1.27 cm (1/2 in.) thick, age-hardened	235	7,060

With this rate of fume generation, proper ventilation is of extreme importance.

In a lecture at the Candian Welding Society's seminar on "Health and Safety in Welding" in 1977, Magnusson offered an excellent overview on fume generation. Of special importance is the equation he used to calculate the nominal hygenic air requirement (NHL), which is the extent of the fresh air dilution (or ventilation) required to render welding fumes harmless. The equation is as follows:

$$NHL = 10 \cdot G \cdot \left(\frac{D_{im}}{5} + \frac{Mn}{2.5} + \frac{F}{2.5} + \frac{X}{x} + \frac{Y}{y} \right) m^3/hr$$

where

G = fume emission in g/hr
D_{im} = inert inorganic fumes indicated in %
Mn = the manganese content in the fumes indicated in %
F = the fluorine content in the fumes, indicated in %
X,Y = The content of all other harmful substances in the fumes in %
x,y = threshold limit values of corresponding substances in mg/m^3

Kobayashi et al. (Ref. 30) measured the amount of fume produced during CO_2 gas welding and found that the CO_2 shield prevented the diffusion of metal vapor and hence decreased the fume generation.

It is known that welding fume composition varies, among other factors, according to the nature of the workpiece. At the same time, fumes generated in welding stainless steel are more active biologically than those generated in welding mild steel. To characterize the particles in these fumes, Eskilsson et al. (Ref. 41) used a computer controlled scanning electron microscope with facilities for x-ray and image analysis. They found that the most biologically active fumes contained particles rich in chromium and associated with potassium.

In view of the possible hazards to the health of workers doing the welding, it was felt by the British Standard Institution that some measure of standardization regarding sampling and analyses of welding fumes should be intro duced. A draft of various procedures for sampling partic ulate matter in the breathing zone of welders, as well a procedures for the various analyses, was submitted i 1977 (British Standard Institute). A similar review wa prepared for the U.S. Environmental Protection Agency by Zimmerman et al. (Ref. 58), in which methods fo the generation and characterization of metal aerosol (mainly for laboratory studies) were summarized.

Gases

Metal fume is not the only source of hazard to th welder's health. In fact, gases that are present in th welder's atmosphere could be more dangerous tha fumes. The main sources of gases are:

(1) Shielding gases; e.g., argon or carbon dioxide

(2) Atmospheric gases; where ozone and nitroge oxides are formed due to the actinic radiation, electri current, and extensive heat

(3) Metal or electrode coatings or degreasing sol vents on metal surfaces; e.g., phosgene

The following gases will be discussed: nitrogen oxide ozone, phosgene, phosphine, and carbon monoxide.

Nitrogen Oxides

The term nitrogen oxides covers the following com pounds:

(1) Nitrous oxide (N_2O), also known as nitroger monoxide

(2) Nitric oxide (NO), also known as mononitroge monoxide

(3) Nitrogen dioxide (NO_2)

(4) Nitrogen trioxide (N_2O_3), also known a nitrous anhydride

(5) Nitrogen tetroxide (N_2O_4), also known a dinitrogen tetroxide

(6) Nitrogen pentoxide (N_2O_5), also known a nitric anhydride.

Nitrogen oxides are evolved from green silage when stored in large silos and hence the name "silo-filler's disease" is used to designate the syndrome that starts with bronchial irritation and culminates in bronchiolitis fibrosa obliterans. Nitrous and nitric acids are other hazardous compounds that are formed when nitrogen dioxide comes in contact with water.

Nitrogen oxides are an irritant to the eyes, mucous membranes, and lungs when inhaled. Exposure to high concentrations of nitrogen oxides may cause severe pulmonary irritation (mainly due to NO_2) and methemoglobinemia (due to NO). The most dangerous effect of nitrogen oxides is the induction of pulmonary edema, which is manifested by cyanosis, cough, dyspnea, chills, fever, headache, nausea, and vomiting or even death.

Chronic exposure may affect the lung mechanics, resulting in a decrease in lung compliance, maximum breathing capacity, and vital capacity, and in an increase in the residual volume. Patients usually show the following signs on medical examination: wheezes and moist rales, sporadic cough, and mucopurulent expectoration. Laboratory testing usually shows an increase in the urinary level of hydroxyproline and acid mucopolysaccharides.

The biological effects of nitorgen oxides, as well as methods for their detection and measurement, are summarized by Fishbein (Ref. 20).

Ozone

Ozone (O_3) represents a higher energy form of atmospheric oxygen. It occurs naturally in the stratosphere as a result of solar radiation and electrical storms. It acts as a filter for ultraviolet radiations, thus protecting humans as well as animals from its harmful effects. The last decade has seen several arguments over the "ozone-destroying theory" of fluorocarbon propellants.

Ozone possesses a characteristically pungent odor, which resembles that of freshly mown grass, clover, or rotting hay, and hence its name; i.e., Zein = to smell (Gr). It is believed that ozone is quite unstable and decomposes rapidly to oxygen, having a half-life of a few minutes to 3 days (Ref. 59). Ozone formed during welding is very readily diluted by other gases and by the intense thermal updraft during welding.

Maximum ozone production during arc welding was reported to occur at wavelengths of 1200 to 2000 Å (Ref. 59). Studying the working environment in the welding and cutting of aluminum, Opekar and Kubovec (Ref. 60) found that the highest ozone concentration in the breathing zone of welders (using GMAW) ranged from 0.2 to 1.2 mg/m^3. The detection of ozone by chemiluminescent methods was reported by Baum (Ref. 61).

Ozone is highly irritating to the eyes and mucous membranes and, in humans, it causes dryness of the upper respiratory passages, coughing, substernal soreness, and, in severe cases, pulmonary edema. Chronic exposure to ozone in experimental animals resulted in the acceleration of lung tumorigenesis in susceptible mice. It is also believed that ozone, due to its free-radical structure, could produce chromosomal aberrations (Ref. 62). Because of its dangerous effects, the TLV for exposure to ozone was set at 0.1 ppm (0.2 mg/m^3).

To study the comparative hazards posed by gases and fumes generated in welding with covered electrodes, Press (Ref. 63) determined the concentrations of ozone and nitric oxide as well as fumes in unventilated closed rooms that were 34 and 121 m^3 in size. He found that the concentrations of ozone and nitric oxide are not dangerous but that the real danger is presented by the very high fume concentrations.

Phosgene

Phosgene ($COCl_2$), also known as carbonyl chloride, is a colorless gas with a sweet, unpleasant odor. It can be formed during the welding of surfaces that contain halogenated hydrocarbons, which may be used for degreasing these surfaces.

As in the case of ozone and nitrogen oxides, phosgene is an irritant to the eyes and mucous membranes. Acute exposure to phosgene results in grey-blue cyanosis, respiratory discomfort, and a feeling of suffocation. Death might follow and would be due mainly to respiratory or cardiac failure.

Chronic exposure of animals to phosgene resulted in emphysema, fibrosis, and chronic pneumonitis (Ref. 62).

Phosphine

Phosphine, or hydrogen phosphide (PH_3), is a colorless gas that possesses an odor of decaying fish. Because it ignites at low temperature, its presence in high concentrations in the welder's atmosphere is not documented. However, Tabershaw et al. (Ref. 62) mention welders among those at potential occupational risk. Phosphine may result from welding on metal recently cleaned with an agent containing phosphate or on metal that has been corrosion protected by phosphatizing processes.

Apart from a local irritant effect on the lungs, phosphine produces systemic effects, such as depression of the central nervous system and damage to the liver and other organs. If exposure is high enough, death results from pulmonary edema or cardiac arrest.

It is believed that chronic exposure to small doses of phosphine does not result in toxicity, primarily because it is rapidly metabolized to nontoxic phosphates in the body.

Carbon Monoxide

Carbon monoxide (CO) is a colorless, odorless gas that is formed by the incomplete combustion of carbonaceous materials. Its formation during welding is due to incomplete combustion of the flames when cold surfaces are being welded.

Carbon monoxide toxicity is caused by the formation of carboxyhemoglobin and thus decreases the ability of the blood to transfer oxygen to the various tissues. If the carboxyhemoglobin level reaches 50 percent, unconsciousness occurs. Carbon monoxide also damages the extrapyramidal system and the basal ganglia in the central nervous system. Its effects are especially dangerous in patients with cardiovascular ailments such as coronary artery insufficiency.

Radiations

Electromagnetic radiations are known to be produced from metal arc welding. According to their frequency and/or wavelengths, radiations are grouped into ionizing or nonionizing radiations. The latter group includes visible, ultraviolet, infrared, microwaves, and radio frequencies. Welders are exposed to only three of the nonionizing radiations, the wave lengths and frequencies of which are mentioned below.

	Wave length (m)	Frequency range (Hz)
Visible	4.0 to 7.6 x 10^{-7}	4.0 to 7.5 x 10^{14}
Ultraviolet	1.0 to 4.0 x 10^{-7}	7.5 to 30.0 x 10^{14}
Infrared	7.6 x 10.0 x 10^{-3}	3.0 to 30.0 x 10^{11}

In a study published in 1978 (Ref. 37), Hinrichs found that radiation, as in the case of fumes, differs from one welding process to another. The order of radiation emission was as follows: gas metal arc welding > shielded metal arc welding > gas tungsten arc welding > plasma arc welding. A positive correlation between the welding current and radiations emitted was also found.

A short account of the hazards from each of these radiations in relation to welding follows.

Visible Radiation

Light can induce its effects on humans in two ways: (1) direct effect in which there are biochemical changes in various tissues due to the absorption of light energy; and (2) indirect effect in which light affects the release of "internal messengers" in the body that regulate, for example, sleep, physical activity, food consumption, and melatonin synthesis.

Welders are exposed to a rather unusual amount of visible light, which might result in photochemical or thermal injury to the retina, or skin burn (ray burn). To protect welder's eyes from such an effect, several shades are recommended and should be regularly used (see Chapter 2 for details).

Zaborski (Ref. 38) attributed the formation of visible light in arc welding to atomic and molecular radiations from vapors and gases in the arc. Visible light produced during welding was also found to be predominantly in the shortwave (420-430 nm) spectrum.

Ultraviolet Radiation

Welders are exposed to ultraviolet radiation produced by arcs operating at high temperatures. The eyes and skin are particularly vulnerable to the injurious effects of UV radiations. The severity of injury depends upon several factors, such as duration of exposure, intensity of radiation, and the presence of sensitizing agents.

Exposure to UV for a sufficient length of time can cause conjunctivitis, which is also known as "welder's flash." Higher exposure levels might result in photokeratitis. Details are provided in Chapter 2.

Furocoumarines and psoralens are photosensitizing chemicals that are present in numerous drugs as well as some plants; simultaneous exposure to UV radiation and these chemicals might produce some skin reactions in man. Examples of toxic effects of UV radiation are given in the following table abstracted from Moss et al. (Ref. 64).

Condition	Wavelength range (nm)	Maximum reaction (nm)
UV carcinogenesis	290-320	290-310
Colar urticaria	290-320 320-400 400-600	varying
Porphyria photosensitivity	380-600	400-410
Xeroderma pigmentosum	290-320	293-307
Polymorphic photodermatitis	290-320 320-400	290-320
Lupus erythematosus (LE) and discoid LE	290-320	
Solar (actinic) degeneration	290-400	?
Photoallergic reactions to halogenated salicylanilides and other related compounds	320-380	330-360
Phototoxic reactions to drugs	320-400 290-320	320-400
Psoralen (8-methoxy- and trimethylpsoralen)	320-280	380-360

In a recent article (Ref. 65), Moss compared the transmission of radiation through 25 transparent welding curtains and found that most transparent curtains will offer protection against UV but not against infrared radiation.

According to Ingram and Horstman (Ref. 66), who studied the near UV radiations (300-400 nm) from welding in a steel factory, the inner shield process is the most intense, with a strong peak at 380 nm.

Infrared Radiation

Infrared emission occurs from various objects having temperatures above absolute zero; the higher the temperature, the more infrared radiation is produced. Absorbed by various tissues, infrared radiation causes an increase in the temperature of these tissues that ultimately leads to the injurious effect.

The most deleterious effect of IR radiation is manifested in the formation of cataracts if the individual is exposed to IR for long enough periods. It takes about 10 to 15 years to develop cataracts in some patients. It was found that wave lengths below 1400 nm are mainly responsible for the development of cataracts.

In certain types of ferrous metal welding, Kuzina (Ref. 67) found that the heat emitted was 1.6 to 6.5 cal/cm^2/min.

Noise

Although the frequency range of audible sounds for human ears is 20 to 20,000 Hz, the range of human speech for regular conversation is 250 to 3000 Hz. This is, therefore, considered to be the most important range, since hearing loss in this range could lead to social handicap. Exposure to intense noise, such as that encountered in welding, could lead to temporary and/or permanent hearing loss in the 4000 to 6000 Hz range. The temporary effect (also known as auditory fatigue) is reversible after spending some time away from the noise source. It is caused by exposure to industrial noises above the 90 dB (A-scale) levels.

Permanent loss of hearing, on the other hand, is irreversible and could be due to several factors other than noise, such as drugs, diseases, and mechanical injury.

Even if noise is not intense enough to cause hearing damage, it is desirable to minimize the amount of noise simply because it can disrupt speech communications, and this affects the welder's performance.

The permissible 8-hour noise exposure level established by OSHA is 90 dB. In a report by Johnson and Shipps (Ref. 68) on health hazards at Corpus Christi Army Depot and in one by Johnson and Roberts (Ref. 69) at the General Motors assembly division in Ohio, excessive noise appeared to be the major health hazard.

Chapter 2

Effects of Welding on Human Health

Advances in industrial hygiene and safety engineering have reduced the risk and, hence, the number of injuries and fatalities among welders. Neglect of personal hygiene and in the use of protecting devices could lead to serious consequences.

In recent years, several reviews and comments on the health hazards of welding have been published (Refs. 1, 70, 71, 72, and 73). One such article deserves special attention (Ref. 39). In this review, Tierney analyzed injuries associated with maintenance and repair in metal and nonmetal mines. To compare and evaluate the hazard potential involved in the use of different equipment, he calculated the hazard index of various functions (hazard index = hazard factor/10^6 man-hours). Of all mining, milling repair, and maintenance types of work, welding and flame cutting were especially high on the hazard index list. His results were as follows:

Function	Hazard index
Welding	5.76 ± 0.26
Burning, cutting	4.96 ± 0.24
Assisting or positioning	2.82 ± 0.23
In vicinity, unprotected	1.70 ± 0.13
Inspecting, cleaning	1.37 ± 0.11

Why does welding appear to be such a hazardous occupation? The reasons appear to be threefold:

(1) There is a multiplicity of factors that can endanger the health of the welder, such as heat or cold (Ref. 74), burns, radiation, noise, fumes, gases, and even electrocution; not to mention uncomfortable postures involved in the work.

(2) The surprisingly wide range in the chemical composition of welding fumes, which vary according to the workpiece, the method employed, and the surrounding environment. The surrounding environment, particularly for maintenance welders, can be a source of exposures to various toxic substances.

(3) The routes of entry through which these harmful agents obtain access to the body.

Actually, hazards due to welding fumes constitute only a fraction of the total occupational risk, when compared to mechanical or accidental risks. In the article mentioned above, Tierney gave the following hazard index for the different actions in welding:

Action	Hazard index
No hood or inadequate hood protection	5.43 ± 0.53
Sputtering or delayed explosion	4.22 ± 0.23
Improper work position	1.88 ± 0.16
Slipping of tool or object	0.89 ± 0.14
Toxic fumes	0.64

Fortunately, hazards associated with the first four situations can be easily overcome if welders exercise caution and alertness. The object of this report is to explore the effects associated with toxic fumes, the 0.64 hazard index.

Background

Vital statistics published by the Registrar General in the UK Health and Safety Executive Guidance Note (Ref. 75) indicate that the death rate in welders is relatively high (SMR 122), especially for diseases such as chronic bronchitis (SMR 109) and pneumonia (SMR 157). Welders are also susceptible to cataracts, deafness, anoxia, pneumoconiosis, heat exhaustion, vibration-induced white finger, and burns.

The bronchitis mentioned above can be caused by the entrance into the body of various noxious compounds, including tobacco smoke. The lungs are the main route of entry for gaseous compounds or aerosols that are readily inhaled. The aerodynamic size of particles in the fumes determines the fate of such invaders. Larger particles are usually deposited on the mucous membrane of the bronchi, where they will eventually be expelled toward the mouth by ciliary movements. Heavy smokers are reported to have lost this defense mechanism. Particles expelled in this manner from the lungs find their way to the gastrointestinal tract. Contamination of food or drink with welding materials constitutes another way by which these compounds can reach the digestive system. Skin represents another possibility, however remote, for the entrance of these compounds into the body. Intact skin constitutes a good barrier against these intruders, but cuts or bruises, or the use of external agents such as degreasing solvents, facilitate the access of various compounds to the body.

These extraneous compounds consist of a wide variety of metals, metal oxides, and organic as well as inorganic compounds.

Toxicity to Various Organs

The effects of welding fumes on the various physiological systems will be discussed.

Effects on the Respiratory System

The surface area of the lung in adult human beings is about 50 to 100 square meters. This constitutes the largest surface area of the body that is potentially exposed to a hostile environment. There are various defense mechanisms by which the lungs protect the body from the effects of noxious substances in the environment. If these defense mechanisms fail to cope with a toxic or hazardous invader, the patient succumbs to respiratory or systemic diseases.

To fully appreciate the "battle" between the welder's lungs and potential invaders, a brief discussion of the nature of the invader and of the lung's defense mechanisms is warranted.

The Invader

For hazardous materials to enter the lungs, they must be in a gaseous form or in a finely subdivided state (liquid or solid) suspended in air (aerosols).

Gases

Gases that are present in the welding atmosphere may:

(1) be produced by the effect of intensive heat or radiation on atmospheric air, e.g., nitrogen oxides and ozone.

(2) arise from shielding gases, e.g., carbon dioxide and argon.

(3) be formed from other sources such as coatings, or electrodes.

Gases act primarily on the lung (ozone, nitrogen oxides, phosgene, phosphene) or elsewhere (carbon monoxide and dioxide).

There are four principal ways in which gases, in general, can be harmful:

(1) They can act as local irritants to the bronchial mucosa, bringing about an increase in mucus secretion and/or bronchospasm. Examples of irritant gases are ozone, nitrogen oxides, and phosgene. Long duration exposure to small concentrations of these gases can result in chronic bronchitis.

(2) Gases affect the availability of oxygen in the atmosphere. This occurs when inert gases such as argon are used in confined spaces.

(3) Gases that chemically bond with hemoglobin affect the ability of hemoglobin to transfer oxygen to tissues, thus resulting in hypoxia or even anoxia. An example of such a gas is carbon monoxide, which is usually present in different degrees in any combustion process. It is also produced from the decomposition of plastic and other coating materials.

(4) Some toxic gases, such as hydrogen cyanide, are carried by blood to certain "target" organs where they exert their systemic effects. A detailed account of these gases was provided earlier (see Chapter 1).

Fine Particles Suspended in Air (Fumes)

The presence of fumes in the welding atmosphere is due to the activity of welding itself (generated from the workpiece or welding wires and rods) or to nonwelding processes such as grinding and chipping. Unlike gases, the toxicity of fine particle solid suspensions in air depends upon the distance they travel in the respiratory system; this in turn is dependent upon the particle size. Usually particles that are less than 0.5 micrometers in diameter obey laws that govern gases. A high percentage of particles that range in size from 0.5 to 0.7 micrometers could reach the alveoli,* whereas larger particles (7 to 20 micrometers) are mainly trapped in the upper respiratory passages. Particles larger in diameter than 20 micrometers are completely trapped in the nasopharyngeal area. Such particles might be hazardous through pulmonary irritation or systemic toxicity (lead, fluorides, chromium, nickel, zinc, vanadium, titanium, magnesium,

*Clapp and Owen (Ref. 76) found that an appreciable amount of welding fumes is in the 1 to 7 micrometer size.

and manganese) or simply by induction of pneumoconiosis. Particulates that induce pneumoconiosis could be relatively harmless (iron, aluminum, carbon), or could cause fibrosis (silica, copper). For asbestos to exert its carcinogenic effect, the material must be inhaled as a fiber whose length is 5 micrometers or more. Such particles could not be identified in welding fumes (Ref. 77).

The Lung

The respiratory passages represent the first line of defense against particulate matters that obtain access to the body by inhalation. The sites of deposition, retention, or clearance; the fate of solid or liquid particulates; and ultimately their toxicologic effects depend, among other factors, on several physical properties such as size, surface area, solubility, and hygroscopicity of these particles.

Various defense mechanisms are available to the body for the expulsion of undesirable substances that reach the lung. The details of these mechanisms are discussed below.

(1) Kratschmer reflex: Exposure of the nasopharyngeal area to irritant gases brings about temporary cessation (stop) of respiration (apnea) to prevent the irritant chemical from reaching the lungs. Changes in the heart rate and blood pressure also occur.

(2) Cough reflex: If the chemical succeeds in reaching the tracheobronchial mucosa, an expiratory blast to expel the irritant vapor occurs.

If the chemical is not an irritant or if it succeeds in reaching deeper into the lungs, there are four additional major defense mechanisms by which the lung deals with the inhaled particulates. These mechanisms are: (1) expulsion by ciliary movement, (2) phagocytosis and lymphatic drainage, (3) clearance by direct intracellular penetration, and (4) solubilization and leaching.

Clearance by Ciliary Movements

The upward movements of the cilia that line the respiratory tract all the way from the terminal bronchioles to the throat result in the clearance of particles deposited in the tracheobroncheal compartment. These particles are then swallowed into the gastrointestinal tract, where they are subjected to various acidic or alkaline media that might result in their solubilization. Insoluble particles are simply excreted. Inhibition of the cilia action by habitual smokers defeats this defense.

Phagocytosis and Lymphatic Drainage

Particles that succeed in reaching the lung are removed mainly by phagocytosis. Phagocytes are then drained into the lymphatic system to the lymph nodes. This process clears the lungs of about 10 percent of the insoluble dust burden.

Clearance by Direct Intracellular Penetration

The degree of solubility of particles in the respiratory free fluids, as well as a particle's size and shape and any biological response that might be provoked by such foreign bodies, are among the various factors that determine intracellular penetration.

Solubilization and Leaching

The speed at which particles that are soluble in the fluids of the respiratory system are cleared from the lung depends upon their solubility rate. Particles that are insoluble in the lung fluids, such as asbestos, are subjected to a process of "leaching" in which any soluble components are gradually extracted by the lungs. The clearance time of certain inorganic compounds, which is expressed as biologic half life, was tabulated by the Task Group on Lung Dynamics (Ref. 78).

These defense processes vary from one individual to the other. Other factors, such as age, drinking and smoking habits, and race, play an important role in determining the extent of toxicity. By dissolving or leaching, some compounds reach the systemic circulation and, eventually, target organs, where they exert toxic effects. There seems to be a certain degree of selectivity in the effect on different organs. For instance, while manganese affects the central nervous system, cadmium seems to concentrate mainly in the kidneys. On the other hand, certain metals such as zinc are capable of inducing generalized malaise (metal fume fever).

Acute Diseases Due to Occupational Exposure

A multiplicity of occupational diseases of the lungs was cited in an article by Wolf (Ref. 79). He attributed a variety of diseases to the noxious gases, aerosols, and fumes that prevail in the welding environment. Furthermore, despite strict adherence to the emission standards in East Germany, Werner (Ref. 80) found a strikingly high incidence of chronic inflammation of the upper respiratory tract, with partially severe course, in welders.

To discuss their effects on the welder's lungs, it is helpful to group gases and fumes according to their action. There are two main groups:

(1) Agents that produce pneumoconiosis (Fe, Al, Cu, Be, etc.)

(2) Agents that produce pulmonary irritation with or without systemic effects

(a) Gases (nitrogen oxides, ozone, phosgene, etc).

(b) Fumes (Cd, Cr, F, Pb, Mn, Zn, Ni, etc.)

Agents That Produce Pneumoconiosis

Pneumoconiosis, a condition characterized by the deposition in the lungs of a substantial amount of particulate matter (dust), has been observed in welders since World War I. The industrial age produced a high incidence of pneumoconiosis, the occurrence of which was linked to mining. As early as 1545, Agricola published the following:

> "The mines are very dry and the constant dust enters the blood and lungs producing the difficulty of breathing which the Greeks call asthma. When the dust is corrosive, it ulcerates the lungs and produces consumption; hence it is in the Carpathian Mountains

that there are women who have married seven
husbands, all of whom this dreadful disease
has brought to a grave."

Of all the diseases that are known collectively as pneumoconiosis, Hanson and Kasik (Ref. 81) singled out siderosis to be the most common disease plaguing welders.

Using advanced scanning electron microscopy (SEM), Guidotti and DeNee (Ref. 40) were able to study lung pneumoconiosis in tissues obtained from a 58-year old arc welder who had been in that occupation for 39 years. Utilizing various modes of SEM, such as secondary electron imaging, backscattered electron imaging, and microprobe analysis, the researchers managed to detect intracellular metal particles that were not otherwise visible. They concluded that "Arc welder's pneumoconiosis appears to be more than a benign siderosis resulting from particulate iron deposition." They maintained that "simultaneous exposure to other components of welding fumes may alter the pathologic picture, inducing a more complicated fibrotic reaction."

Stettler et al. (Ref. 43), using x-ray analysis of lung biopsy specimens from two welders, showed that the majority of particles consisted of iron, chromium, manganese, and nickel. The percentage of various metals in both cases was as follows:

Metal	Case 1	Case 2
Stainless steel	81.6	68.8
Aluminum	3.4	21.8
Silicate	6.2	4.8
Silica	5.0	2.4
Other	3.8	2.2

Microscopic examination of lung tissues showed extensive interstitial fibrosis in the second case, but not in the first. Examination of the above table reveals that the only significant difference in the composition of these metals is the higher percentage of aluminum in the second case. Whether this is the cause of fibrosis could not be determined.

Whether or not iron oxide brings about interstitial fibrosis of the lungs is a matter of controversy. Patel et al. (Ref. 42) reported on a case of acute bronchitis with interstitial pneumonitis and fibrosis in a patient who had spent 18 years as an arc welder in a Navy dockyard. Prior to his admission, the patient complained of exertional dyspnea and productive cough for seven years. Pulmonary function tests showed moderate restrictive disability with some reversible obstruction. Radiographic examination revealed diffuse fine reticular opacities, especially on the right side. No details were given about the nature of this patient's exposure to welding fumes, nor was the treatment mentioned. The researchers concluded that iron oxide is fibrogenic and that fibrosis might be intensified by simultaneous exposure to asbestos. This opinion is shared by Fabre et al. (Ref. 82).

Siderosis of the lung is not the only finding in lung biopsies of 14 welders studied by Irmscher et al. (Ref. 83). In fact, the authors report detecting lung cirrhosis in some cases and attributed it to welding fumes.

Roussel and Pottier (Ref. 84) reported on a case of pulmonary berylliosis in an arc welder who spent about 15 years using varying welding processes in a beryllium plant. The disease is characterized by thoracic pain, dyspnea, and radiographic changes. These symptoms are reversible, according to Sprince et al. (Ref. 85). In a survey of workers in a beryllium extraction plant, they found that of 31 workers with interstitial disease, 20 suffered from hypoxemia, and 11 from a combination of hypoxemia and radiographic changes. At that time, the concentration of beryllium in air was found to be 1290 $\mu g/m^3$, which is more than 50 times the accepted exposure limit (25 $\mu g/m^3$). Improvements in engineering and ventilation of the plant led to a sharp decrease in the air concentration of beryllium (15 $\mu g/m^3$). Three years later, a follow-up study showed that only 13 patients showed hypoxemia, and that radiographic changes were reversed in 9 out of 18 workers who remained on the job.

Agents That Produce Pulmonary Irritation

Both gases and fumes can produce pulmonary irritation, with or without systemic effect.

Gases

The gases discussed are nitrogen oxides, ozone, phosgene, phosphene, and hydrogen fluoride.

Nitrogen Oxides. Nitrogen dioxide (NO_2) and its dimer nitrogen tetroxide (N_2O_4) are collectively known as "oxides of nitrogen." This term is a misnomer and should not be confused with nitrous oxide gas that is used as an anesthetic. Nitrogen oxides NO, NO_2, N_2O_3, N_2O_4, are formed by the combination of atmospheric oxygen and nitrogen under the heat of the electric arc. If ventilation is adequate, it is unlikely that nitrogen oxides will accumulate to a degree exceeding a TLV of 5 ppm. In confined spaces and especially when an oxygas flame is used, dangerous concentrations can be reached in as short a period as 15 minutes. Nitrogen dioxide (NO_2), the most toxic of these gases, produces its injurious effect by combining with water in tissues to form nitrous and nitric acid (Ref. 86). The first reported death from nitrous fumes dates back to 1804. Exposure of humans to nitrous fumes occurs in occupations other than welding, such as in the chemical munitions, and missile industries. After exposure, there is usually a delay of from 3 to 30 hours before symptoms of lung irritation appear. The symptoms include a dry, irritating cough, pain in the chest, shortening of breath, and cyanosis. Later, pulmonary edema and frothy sputum, often canary yellow in color, invariably appear. Usually this is the most critical period when most patients succumb. However, those who survive are apt to suffer a relapse two to six weeks later with the clinical features of bronchiolitis obliterans.

In an excellent review of the toxicity of nitrous fumes,

Prowse (Ref. 86) concluded that chronic exposure to small concentrations of these fumes may cause emphysema. The author also concluded that "permanent damage may result at lower concentrations than" the TLV of 5 ppm.

Kurta (Ref. 87) reported on a case of acute pulmonary edema in a 47-year old New Zealand man, who was welding in a confined space (5000-gallon aluminum tank). The metal had been preheated using oxyacetylene flame. The welding process then continued with the tungsten arc argon shielded method. Pulmonary edema developed about five hours after completion of welding, and it was suspected that the patient had been exposed to a mixture of nitrogen oxides and ozone. Treatment with ampicillin and prednisone brought about rapid remission of the symptoms with complete recovery in 11 days.

Another case involved four Korean welders, reported by Hocking (Ref. 88). Although no details were given, these patients apparently suffered from nitrous fume toxicity. This is evident from their complaint of cough with yellow sputum, retrosternal pain, and mild shortness of breath following the use of electric arc welding 600 feet underground. The 1000-foot long ventilation tunnel was partially blocked by falling rocks. Symptomatic treatment was given.

Ozone. Ozone is produced by ultraviolet radiation in the air in the vicinity of arc welding. Although it is produced in small amounts in arc welding with coated rods, significant amounts are generated in gas shielded welding processes. Ozone is a pulmonary irritant gas and could cause bronchitis and/or pneumonia.

Phosgene. A conference on the long-term effects of halogenated solvents agreed that welding in an atmosphere containing halogenated hydrocarbons might produce toxic concentrations of phosgene due to the thermal and photochemical decomposition of such solvents as trichloroethylene, perchloroethylene, and methyl chloroform (Ref. 89). Phosgene has a delayed action on the lungs similar to nitrogen oxides and ozone. Because of its characteristic odor, phosgene could give an early warning of its presence in the welding atmosphere.

Fumes

Welders who worked in a machine plant assembly room for more than 10 years were examined by Gola and Galazka (Ref. 46). Of the 73 welders examined, 60 percent complained of coughs, expectoration, dyspnea, and frequent acute rhinitis. Clinical examination revealed that 10 percent of the welders showed symptoms of bronchitis, pulmonary emphysema, and lung adhesions. Only one case of fibronodular tuberculosis was found. Cases in which pulmonary diseases were attributed to certain components were as follows:

Cadmium. Exposure to cadmium fumes causes pulmonary edema. Wolf (Ref. 44) reported on a case involving five men who were exposed to cadmium fumes through heating cadmium-coated metal with an oxyacetylene torch in a confined space. One of the victims died within five days from massive pulmonary edema and thrombosis as well as extensive kidney damage.

Stanescu et al. (Ref. 6) examined 18 workers who were exposed to cadmium for an average of 32 years. Results showed that, although these workers had a significantly higher level of cadmium in their blood when compared to a control group of 20 nonexposed workers, only grade 1 dyspnea was observed. The authors concluded that cadmium does not induce emphysema.

Chromium. Mosekilde and Pallisgaard (Ref. 90) examined 18 patients with idiopathic alveolitis fibrosa to determine to what extent their pulmonary disease was caused by exogenous agents. In only one patient was the ailment ascribed to the inhalation of chrome fumes. That patient developed chrome-eczema simultaneously with the pulmonary symptoms.

To study the effects of chromium, Jindrichova (Ref. 29) compared three groups of welders. The first group, averaging 45.4 years in age, included nine welders who worked mainly with basic electrodes under what she termed "very good" hygienic conditions. The second group, with an average age of 40.7 years, included 11 welders who used basic electrodes for half of their working time; in the other half, electrodes with up to 23 percent chromium content were employed. The third group of 11 welders spent three months prior to a medical examination using electrodes rich in chromium for more than 70 percent of their working time. No adverse effect was found in the first group. However, 36 percent of the second group and 72 percent of the third manifested chronic bronchitis. Furthermore, only in group three did the author find erosion of the nasal septum (among 36 percent of the welders). Also, chromium excretion in urine was significantly higher in group three than in either the control or the other two groups.

Lung Function Test

The use of the lung function test alone as an indication for occupational lung damage is dangerous, since lung function tests may appear normal despite other manifestations of disease. This is especially true in cases with small radiological lesions, the so-called simple pneumoconiosis. Some functional indices reflect simple pneumoconiosis, but these indices alone may be misleading without other thorough studies. Not until pneumoconiosis is complicated by chronic bronchitis or emphysema are serious derangements in respiratory function detected.

To study the correlation between the degree of lung damage and changes in lung function test, Marek and Kujawska (Ref. 91) examined three groups of patients with silicosis, coal worker's pneumoconiosis, and arc welder's pneumoconiosis. The same tests, viz., lung radiograms and lung function tests, were repeated six years later. The authors' findings: "Functional disorders do not parallel radiographic changes." They also added that, in arc welder's pneumoconiosis, the distinct regression of the radiographic changes is an indication that this type of pneumoconiosis is not collagenous.

In a study published in 1977, Spacilova and co-workers (Ref. 92) assessed the pulmonary changes among 42 Czechoslovakian arc welders. The lung function and radiograms of 22 welders, who were selected on the basis of not having chronic bronchitis, were compared to those of 20 control men. Although there was more evident radiographic change in 50 percent of the welders in terms of size and profusion of lung opacities, no significant difference from the control group was observed insofar as lung function was concerned.

In a study of 62 welders, Bergert et al. (Ref. 93) found that workers who are more than 40 years of age, with at least 20 years of active work, exhibited signs of chronic obstructive diseases of the respiratory tract as reflected by lung function tests. The authors concluded that these effects are due to welding fumes and gases.

Lung function tests, including closing volume (CV), closing capacity (CC), total lung capacity (TLC), and the slope of the alveolar plateau were determined by Oxhoj et al. (Ref. 94). That study, which was conducted on 119 Swedish welders and 90 unexposed controls, showed that CV, CC, and TLC were significantly lower in welders who were nonsmokers and were exposed for at least five years to welding fumes.

Lung Clearance

Kalliomaki et al. (Refs. 95 and 96) measured the amount of iron retained in the lungs of 42 shipyard arc welders who were exposed to welding fumes for a period of one to 40 years, using a method that was developed earlier (Ref. 97). They found that the amount of alveolar deposition averaged 20 to 40 mg per year and that the clearance rate was 10 to 20 percent per year.

One of the primary concerns in welding is to protect the eyes, face, and neck from the effects of fumes, radiation, heat, and flying sparks.

Effect of Fumes

Alkali and acid fumes have a deleterious effect on the eye and can cause serious ocular damage. By reacting with lipids in the corneal epithelial cells, alkali substances form a soluble soap that can quickly penetrate the corneal stroma and the monolayer of endothelial cells and enter the anterior chamber of the eye. In most cases, the damage to the corneal epithelial and stromal cells is irreversible. The process of repair and regeneration of viable cells in these damaged tissues is extremely slow if not impossible. Degeneration and sloughing of the tissue of the anterior segment might end in perforation of the globe and loss of the eye.

Acid fumes, on the other hand, are generally more tolerable than alkali fumes. This is mainly because the eye has a marked capacity for neutralizing acids, as do most other tissues.

Effect of Radiation

The heat and radiation usually produced during arc welding pose another hazard to the welder's eyes. Short wavelengths (200 to 315 nm) are absorbed by the nucleoproteins in the corneal epithelial cells. Continuous exposure to these wavelengths leads to abiotic changes and fragmentation of the nuclei and the ultimate death of the epithelial cells. This leads to the exposure of the bare nerve endings that are present in the cornea as a fine network surrounding the epithelial cells. Contact of these nerve endings with tear film and the mechanical effect of eyelid movements produces such intense pain that patients cannot open their eyes. This condition is called actinic keratitis or photo-ophthalmia (Ref. 98). Essentially three types of radiation are involved: visible light, ultraviolet, and infrared.

Injuries Due to Exposure to Visible Light

Because of the use of protective glasses, retinal burns due to exposure to excessive luminance have not hitherto been reported. Also, discomfort from visible light occurs at approximately 1 percent of the illumination level required to induce retinal injury.

The inability of the eye to adapt to excessive light is a common phenomenon. Zaborski (Ref. 99) found that the difference between the arc luminosity and that of the surrounding area is the factor mainly responsible for this effect. It is similar to emerging from a dark restaurant in the brilliant noon sunlight. Results of permeability indices of various glass screens were also presented in this study.

Krasniuk and co-workers (Ref. 100) reported that high ampere argon arc welding has a deleterious effect on the hemo- and hydrodynamics of the eyeball.

Injuries Due to Ultraviolet Radiation

Exposure to ultraviolet radiation affects the corneal epithelium. According to Tengroth and Vulcan (Ref. 101), 7000 eye incidents were reported in one year, of which about 30 percent were caused by exposure to ultraviolet radiation. This fact prompted NIOSH to establish a permissible dose for an eight-hour period (Ref. 102). These doses are summarized in the following table, which shows that the wavelength of 270 nm is the most harmful to human eyes.

Permissible 8-hour doses of ultraviolet exposure (Ref. 102).

Wavelength (nm)	Permissible 8-hour dose (mj cm^{-2})
200	100.0
210	40.0
220	25.0
230	16.0
240	10.0
250	7.0
254	6.0
260	4.6
270	3.0
280	3.4
290	4.7
300	10.0
305	50.0
310	200.0
315	1000.0

Excessive exposure to ultraviolet radiation causes what is known as "arc welder's flash," characterized by conjunctivitis and keratitis. Ultraviolet radiations can also damage the lens if the dose that affects the cornea is increased threefold. Proper filter plates should be used to protect the eyes from these emissions. Filter plates recognized by the National Bureau of Standards are identified by shade number. The following shade numbers are recommended for various welding operations (Ref. 103).

Arc welding operation	Suggested shade number
Shielded metal arc welding, up to 5/32-in. electrodes	10
Shielded metal arc welding, 3/16 to 1/4-in. electrodes	12
Shielded metal arc welding, over 1/4-in. electrodes	14
Gas metal arc welding (nonferrous)	11
Gas metal arc welding (ferrous)	12
Gas tungsten arc welding	12
Atomic hydrogen welding	12
Carbon arc welding	14
Light cutting, up to 1 in.	3 or 4
Medium cutting, 1 to 6 in.	4 or 5
Heavy cutting, over 6 in.	5 or 6

Injuries Due to Infrared Radiation

Mean infrared radiation can reach deep into the eye to affect the retina. Wavelengths beyond 1.9 nm are absorbed by the cornea while those greater than 1.4 nm are absorbed by the cornea and aqueous humor. The absorbed energy raises the temperature of the cornea and of the internal structures of the eye. It is this heat that plays an essential role in the development of opacities in the lenses of welders. Infrared radiation is intense during arc welding and decreases during other ordinary welding processes. Filters having a damping factor of 10 to 10^4 can be used to protect the eye from infrared radiation. Transmittance requirements for filters used in welding are discussed in a technical review by Kasansky (Ref. 104). A project to construct new safety glasses (with an opening of 10 nm) for Swedish welders was discussed by Tengroth (Ref. 105).

Medical examination of 926 welders over a six-year period by Ross (Ref. 44) showed that, contrary to what is believed by welders, welding has no long-term effects on either near and far vision as compared with a control group of 755 persons. Only two cases of cataracts were found, which Ross reported to be of traumatic origin and unrelated to infrared exposure. The incidence of foreign bodies in the eyes was as follows.

Frequency	% Welders
Less than once	13
One or two	16
3-9	10
More than 10	21
Never	41

Effects on the Ear and Hearing

Welders may be exposed to excessive noise, either of their own making or from neighboring workers, during grinding or chipping processes. Plasma arc welding, in which the plasma jet is rushed through the torch nozzle at a very high speed, can result in intense high frequency noise from which welders should be adequately protected.

Because of their exposure to noise for long durations, it is not surprising that many welders lose some hearing capacity. Studies by Ross (Ref. 44) of 926 welders revealed that 33 had an average hearing loss of 34 dB or more in the range of 1 to 3 kHz, which is the speech frequency. Loss of hearing to that degree could lead to a social handicap. Moreover, he noticed that a substantial loss of hearing occurred in the range of 4 kHz.

Effects on the Skin

Although welders are exposed to a variety of hazardous materials, most skin conditions are the result of burns from hot metal. Furthermore, ultraviolet radiation can pass through thin cotton clothing and is capable of inducing ray burn. In view of its meltability and flammability when in contact with hot metals, Sloan (Ref. 73) suggested that nylon clothing should not be used to protect manual arc welders. A classical, though uncommon, injury is the circumferential burn, which is electrothermally induced by the arcing of an electrical current through a finger ring. Crow and McCoy (Ref. 106) reported on the use of the "Z-plasty" technique for the treatment of such conditions.

Jirasek (Ref. 107) reported on ten cases of skin trauma (occupational tattooing) caused by clinkers during spot welding in a car factory.

Effects on the Gastrointestinal Tract

An investigation on the health effects of low hydrogen electrodes in welding was performed by Singh and Gupta (Ref. 45). The composition of the fume generated was as follows:

Content	Mean
Nitrogen dioxide	0.82 ppm
Ozone	0.07 ppm
Carbon monoxide	20 ppm
Iron oxide	15.06 mg/m^3
Fluoride	4.76 mg/m^3
Manganese	0.85 mg/m^3

Among workers exposed to these fumes, five welders complained of loss of appetite and stomach ache.

Effects on the Cardiovascular System

Clinical examination of 73 welders employed in a machine factory for more than 10 years revealed that 8 percent (age 40 to 50 years) showed dynamic arterial hypertension and radiological symptoms of aortosclerosis (Ref. 46).

Some metals in particulate form are known to exert a hemolytic effect on the human erythrocytes. To study the mechanism by which these particulates bring about hemolysis, Rae (Ref. 47) added human erythrocytes to a suspension of silica (5 μm), nickel (1 μm), cobalt (1 μm), chromium (5-30 μm), iron (75 μm), molybdenum (5-6 μm), cadmium (5-60 μm), and zinc in veronal buffer. The most active, i.e., damaging to blood cells (and macrophages), particulates were cobalt and nickel. Their effect was second to silica, which was used as a positive control. Cadmium and zinc were the least damaging particulates. Intermediate effects were observed with chromium, iron, and molybdenum. The author postulated that the following two mechanisms could possibly explain the hemolytic effect of particulates: (1) an immediate and direct effect on the cell membrane (which is observed in a few minutes); and (2) a slow developing effect that results from metal dissolving from particulate material. The most significant effect of this second phase may be the release of lysosomal enzymes, which could result in cell death.

Gola and Galazka (Ref. 46) reported on one case of granulocytopenia among 73 welders who were employed for more than 10 years in a machine factory. The patient was 39 years of age and was employed for 20 years as a welder. A case of leukopenia occurred in a 35-year old welder who worked for 10 years in the same plant.

The formation of carboxyhemoglobin in welders exposed to a low atmospheric concentration of carbon monoxide was studied by Rogowsky and others (Ref. 108). The study was performed on welders working on an automobile assembly line. After six hours of work, nonsmoker welders showed an increase of 1.49 percent in their carboxyhemoglobin level, as compared with a 5.25 percent increase in moderate smokers and 8.43 percent for heavy smokers. Similar findings were reported earlier by Bauer et al. (Ref. 109).

Effects on the Central Nervous System

Three groups of electric welders in various stages of manganese poisoning, viz., subclinical, functional, and organic stages, were subjected to experimental-psychological investigations (Ref. 110). The author found derangement in short-term memory, coordination, and concentration, as well as visual motor disturbances, even in those in the subclinical stages.

Toxic manifestations of manganese were not elicited in a group of 15 workers (1 welder and 14 assembler-grinders) involved in the fabrication of railroad track from an alloy containing about 12 percent elemental manganese. The ambient concentrations of manganese and iron oxides were 0.53 and 3.60 mg/m^3, respectively (Ref. 111).

Effects on the Liver

The toxic effects of metal fumes on the liver are shown as biochemical changes, which are discussed later. Avakyan et al. (Ref. 112) reported on some liver disorders that were observed in workers in the copper-molybdenum industry.

Effect on the Musculoskeletal System

No article was published in the 1978-79 period of updating the previous report on welding.

Effects on the Reproductive System

According to our best knowledge, no article on this subject was published in the 1978-79 period.

Effects on the Urinary System

It is well known that exposure to such heavy metals as cadmium and lead produces damage to the proximal tubules of the kidney. Although pure cadmium is rarely used in the welding industry, it is an important constituent in various alloys. Cadmium is used as a substitute for tin in antifriction metals (Ref. 113).

Exposure to cadmium fumes may occur from welding of cadmium-plated or painted metals. Acute exposure to cadmium fumes causes irritation of the respiratory system as a primary target. However, long-term exposure causes proteinurea characterized by an abundance of low molecular weight proteins.

In a study by Kjellstrom et al. (Ref. 114), the urinary excretion of beta-2-microglobulin of 240 workers (of both sexes) exposed to cadmium oxide and nickel hydroxide dust in a Swedish factory was compared with a control unexposed group of 87 males. Results showed that the increase in the urinary excretion of beta-2-microglobulin correlates with the length of employment. Thus, an average increase of 19 percent was observed in workers with six to 12 years of exposure as compared to 3 percent for the control group. The average level of exposure was 50 μg Cd/m^3. Furthermore, the authors found that smokers had a three times higher rate of excretion than nonsmokers.

The British Occupational Hygiene Society Committee on Hygiene Standards (Ref. 115), assessing the hazards associated with the inhalation of cadmium oxide, concluded that long-term exposure to cadmium dust or fumes leads to permanent lung damage as well as renal tubular dysfunction. A hygiene standard of 0.05 mg cadmium/m^3 for all cadmium compounds was recommended.

Effects on the Endocrine System

No articles on this subject could be located.

Effects on the Teeth and Oral Cavity

Exposure to fluorides in welding fumes presumably can have an undesirable effect on welders' teeth. However, to our knowledge no article was published dealing with this phenomenon.

Metal Fume Fever and Allergic Reactions

Metal fume fever, an acute episode of short duration, is attributed to the inhalation of metal fumes. It was first described by Potissier in 1822 (Ref. 116) and is also known as copper fever, zinc chills, brass chills, brazier's disease, welder's ague, Monday fever, galvanizer's poisoning, and foundry fever.

The phenomenon is caused by such metals as cobalt, aluminum, selenium, cadmium, silver, manganese, iron, antimony, nickel, beryllium, tin, and vanadium; but most notably by magnesium, zinc, and copper. It is characterized by respiratory, gastrointestinal, and central nervous system involvement. Symptoms are of sudden onset and usually begin with a metallic taste in the mouth accompanied by thirst. This is followed by coughing, respiratory distress, muscular pain, headache, polyurea, fever, profuse sweating, and sudden chills. Gastrointestinal involvement, although uncommon, is evidenced by nausea, vomiting, and a foul taste in the mouth. Usually, these signs and symptoms subside in 24 hours and complete recovery ensues (Refs. 117 and 118).

In 1977, Gun (Ref. 119) reported on the frequency of febrile episodes that occurred in 27 welders as compared to a control group of 27 nonwelders. Generally, a significantly higher incidence of febrile episodes was found among the welders (81 percent) than with the control group (44 percent). However, when the author stratified the data according to the duration of febrile episodes, he found that the difference was due mainly to episodes that were sustained for three or more days. Since the duration of a typical metal fume fever is usually one day, he ascribed these findings to an increased susceptibility of welders to other febrile illnesses. However, metal fume fever was found to occur at a noticeably higher rate in welders who smoke (64 percent) when compared to welders who do not smoke (25 percent).

During 1978, two cases of metal fume fever were reported. In the first case (Ref. 120), a 24-year old nonsmoking welder experienced headache, nonproductive cough, and orthopnea 15 minutes after the start of a procedure for melting the following combination of metals: beryllium 2 percent, aluminum 1 percent, and copper 97 percent. The procedure also involved the use of chlorine, nitrogen, alcohol, and adhesive glue. An hour later, the welder felt feverish and complained of substernal pain. Upon admission to the hospital, he had sinus tachycardia and a respiratory rate of 28/minute. Within 30 hours, these symptoms subsided; however, results of pulmonary function tests continued to be below normal, showing a decrease in forced vital capacity (FVC), and in forced expiratory volume (FEV_1) in one second. The ratio of FEV_1/FVC was 66 percent (normal is 75 percent). Serial pulmonary function tests showed persistent mild to moderate obstructive impairment of the lung, even at three months after the welder quit working at the factory. However, other symptoms, such as leukocytosis and hypoxemia, were gradually resolved without any medication.

The second case was reported by Dula (Ref. 121). Unlike the first case, this welder, a 35-year old male, developed chills and dyspnea three hours after he had finished welding. He also worked for two hours longer than usual, using a new metal whose composition he did not recall. Signs and symptoms were much like those in the case described above. No pulmonary function test was performed. However, an x-ray of the lung showed a haziness of the left upper lobe and infiltrate in the left lower lobe. Oxygen was given by a nasal tube and arterial blood gases were monitored. Within 24 hours, the fever had subsided and the chest x-ray was normal.

In his book about the effects of environmental pollutants on health, Waldbott (Ref. 122) devoted an entire chapter to fever-producing agents. Detailed discussions of various metals that cause fever, such as manganese and zinc as well as other fever-inducing pollutants, were included. Various aspects of zinc-induced fever were also discussed by Ferris (Ref. 123).

Calnan in 1979 (Ref. 124) reported on a case of metal fume fever that exhibited, apart from the typical symptoms, some skin involvement in the form of skin rash and eczema.

Treatment of Metal Fume Fever

Simply by removing patients from the polluted environment, symptoms of metal fume fever usually subside within 24 to 48 hours. However, in ceratin cases, such as exposure to cadmium fumes or in prolonged exposure, fatal acute pneumonitis might follow (Ref. 121). Hopper and Dula (Refs. 120 and 121) warned that mistaking metal fume fever symptoms for those of influenza or bacterial pneumonia can be disastrous.

The main lines of treatment of metal fume fever are measures to correct hypoxemia bronchospasm and lung injury. Hypoxemia can be easily treated by the inhalation of oxygen. Alleviation of bronchospasm can be achieved by the administration of a bronchial muscle relaxant, such as aminophylline (Ref. 121). Lung injury can be reduced by the administration of corticosteroids, although this could mask any secondary lung infection; in which case, antibiotics should be used.

Laboratory tests for detecting borderline cases of exposure include chest x-ray examination, aortic blood gases determination, and the use of shunt studies (Ref. 121).

Biochemical Changes

The complexity of assessing health effects of exposure to welding fumes can be appreciated if one considers the numerous and endless variations in the composition of these fumes. Different components of the fumes exert an effect on human biological systems that varies not only in magnitude but also in direction. Thus the net effect may be the algebraic sum of all these effects, as well as any other local substances in the lungs (the release of which might be affected by these metals). For example, exposure to zinc prevented the development of lead poisoning in young horses. This antagonistic effect of lead and zinc on δ-aminolevulinic acid dehydratase (ALA-D) was demonstrated by Cantrell et al. (Ref. 125), who showed that the decrease in ALA-D activity in baboons exposed to lead oxide (1.5 mg/m^3 for 4 hours/day, for 26 days) was reversed by zinc. Exposure to lead is known to bring about a decrease in the activity of erythrocyte δ-aminolevulinic acid dehydratase enzyme (ALA-D). This is a highly sensitive indicator for exposure to lead, especially at low levels.

In a study by Beretta et al. (Ref. 126), the lead-blood level (PbB), delta amino levulinic acid level in urine (ALAU), delta amino levulinic acid dehydrase (ALA-D), and urinary coproporphyrine (CPU) were estimated in a group of 33 women employed in the microwelding of printed circuits with lead-tin alloys. Results showed a higher level of lead in the blood of these workers than in a group of young women who were not occupationally exposed to lead. The authors recommended the use of ALA-D as a suitable biological test for the detection of low lead exposure.

A paper by Graben et al. (1978) titled "Acute Heavy Lead Poisoning in Welders" described intoxication by lead among six workers who used blow torches to demolish a bridge painted with lead paint. They complained of nausea, vomiting, abdominal colic, myalgia (especially in the back muscles), pain in the joints, parasthesia, and sleep disturbances. Kidney involvement was indicated from the pain induced by palpation in the area of the kidneys as well as by a decrease in creatinine clearance. High pathological levels of lead were found in blood (890 to 1000 μg/1) and in urine (43-1740 μg/24 hrs). There was also an increase in the urinary excretion of δ-amino levulinic acid and porphyrins. The authors found no correlation between the level of lead in blood or urine and the severity of the clinical symptoms. Rather, the quantity and pattern of excretion of porphyrins and their precursors are especially valuable for the diagnosis and treatment of lead poisoning. The following pattern of porphyrins excretion was claimed to be characteristic of lead poisoning:
copro- $>$ pentacarboxy- $>$ tricarboxy- $>$ uro-porphyrin.

This paper illustrates some of the confusion in the clinical literature by health specialists. The occupation described was "flame cutter" or "demolition worker," not welder. It was pointed out earlier that lead intoxication in welders has not been reported. The preceding citation illustrates the problem.

The determination of lead, cadmium, and zinc in biological materials using the isotope dilution mass spectrometry was described by Gramlich et al. (Ref. 127).

To determine the extent to which occupational exposure to cadmium and nickel affect health, Thoburn and Larsen (Ref. 26) conducted an evaluation of the health hazard of workers at a battery company in the West. Workers selected were "with potentially high dust and welding fume exposure" (Departments 330A and 340A), workers on the reject table and core rollers (Department 330B), as well as stackers and combers (Department 340B). Analysis of cadmium, nickel, and zinc in hair (samples were taken from the pubic area) and urine of exposed workers as compared to a control group (Department 350) revealed the following results:

Substance	Specimen	Group	No.	Mean values found	Controls (350)
Cadmium	Hair	Control (350)	14		15.2 μg/g
		340B	11	61.5 μg/g	
		330A			
		330B	27	197.2 μg/g	
		340A			
	Urine	Control (350)	9		0.024 μg/mg creatinine
		330 A & B		0.052* μg/mg	
		340 A & B	33	creatinine	
Nickel	Hair	Control (350)	14		6.14 μg/g
		330A, 340A	20	43.38 μg/g	
		330B, 340B	18	21.36 μg/g	
	Urine	Control (350)	9		121 μg/24 hrs
		330A	9	418 μg/24 hrs	
		340B			
		340A	24	141 μg/24 hrs	
		340B			
Zinc	Hair	Control (350)	14		104 μg/g
		All others	47	166 μg/g	
	Urine	Control (350)	9		366 μg/24 hrs
		All others	40	499 μg/24 hrs	

The concentrations of cadmium and nickel in the hair of all exposed workers were significantly higher than those of the control. The cadmium level in the urine of all workers was increased, while that of nickel was increased only in group 330A. No significant changes occurred in the zinc level, either in hair or urine.

Medical examination showed that workers in Departments 330A and 340A had the highest incidence of symptoms indicative of upper respiratory irritation. The authors concluded that this is "probably due to exposure to plate dust and tab welding fumes."

Carcinogenicity of the Fumes

Reported cases of cancer in welders will be discussed according to the organ affected.

Skin Cancer

In a review article by Emmett (Ref. 128), skin cancer as it relates to occupations was discussed. The author maintained that UV radiation and polycyclic aromatic hydrocarbons are the major causes of skin cancer. Radiation-induced skin tumors have a latent period that ranges from seven weeks to almost half a century, depending upon the dose. Welding is one of the occupations in which workers are known to be exposed to ultraviolet radiation.

A team of four surgeons from Gainesville, Florida, (Ref. 48) reported on a case of soft tissue tumor of the thumb that had morphological features of "clear-cell sarcoma of tendous and aponeuroses." The case involved a 59-year old white male who, in 1955, had a traumatic injury of the left thumbnail in which a "black welding flux" was introduced under the nail. Thirteen years later, the patient first noted the presence of a nodule in the dorsum of the distal phalanx of the involved thumb, where several coalescing nodules gradually developed. Pain was noted for the first time just one year prior to his admission to the hospital; x-ray examination revealed osteomyelitis of the distal phalanx. On admission to the hospital, x-ray of the thumb showed destruction of the distal phalanx with probable involvement of the adjacent portion of the middle bone. After a biopsy revealed an epithelioid-cell malignant tumor, a ray amputation of the thumb was carried out. One and one-half years later, a large nodal tumor in the axilla was found; one year later, evidence was found of pulmonary metastasis.

Brain Tumors

In 1975, Jaremin and colleagues (Ref. 17) reported on a brain tumor case in which the symptoms were manifested in the course of zinc fever. The victim was a 52-year old man who had spent the last 22 years of his life as an electric welder; previously, he had been a coal miner. Prior to admission to the hospital, he had been welding zinc-coated pipes and sheets for two weeks, 8 hours per day, in a shipyard where zinc fever had frequently been noted. The symptoms were typical for metal fume fever and consisted of a sudden onset of cough, fever, headache, and chest pain. The symptoms were further accentuated and gradually replaced by drowsiness, failing memory, and unconsciousness. The patient's past history showed welding gas poisonings (unspecified) and pneumonias. Absorptive spectrophotometric analysis of serum-metal levels revealed the following concentrations in microgram percentages*: copper 114; lead 80; cadmium 5; manganese 10; and zinc 580. (The normal values for these metals are: 130, 80, 1.8, 0.6, and 120 microgram percent, respectively.) The content of zinc in the erythrocytes was 2290 μg percent and that in urine was 1500 mg/L. Normal measures to support the circulation and respiration were of no help. The patient was then referred to the neurosurgical unit where cerebral artenographic examination revealed the symptoms of a tumor in the left temporal region. The patient was then operated on and a cyst removed from the left parietal lobe and the third temporal gyrus was resected. After the operation, the patient developed aphasia and died three weeks later. Autopsy showed a malignant osteocycoma, various bronchial effects, and either soot or particles of coal dust in the lungs.

Zinc is normally present in various parts of the brain, such as the cerebellum, hippocampus, and the gray matter. As tumor development occurs over a period of years if not decades, and as the exact nature of the above patient's exposure to unknown metals for 22 years was not mentioned by Jaremin, it is unknown whether exposure to zinc was the sole stimulus that led to the development of brain tumors. Nonetheless, it seems justifiable to assume that excessive zinc absorption might have partially contributed to the unveiling of the latent clinical symptoms of brain tumors.

Cancer of the Lung and Nasal Sinuses

An epidemiological study conducted in England by Doll et al. (Ref. 49) to determine the occupational risks of lung and nasal sinus cancer in nickel refinery workers was published in 1977. Nine hundred and sixty-seven men who were employed for at least five years before 1945 were selected for this retrospective study, which investigated the group's mortality rates from different causes. Of the group, 689 died before January 1, 1972. Data was stratified according to the year of first employment, as follows: before 1910, 1910-14, 1915-19, 1920-24, 1925-29. The risk of death from carcinoma of the nasal sinuses was 300 to 700 times the national average for employees who started work before 1920, and 100 times for those employed between 1920 and 1925. No deaths were attributed to cancer of the nasal sinus in workers who started after 1924. However, one man who started in 1929 developed a cancer of the ethmoid sinus. His death in 1967 at the age of 73 was attributed to heart failure due to aortic stenosis; cancer of the ethmoid was referred to as a contributory cause. The decline in the death rate from cancer of the nasal sinuses was attributed to changes in the design of the plant as well as to the adoption of some hygienic measures, such as the use of cotton pads on the face.

Although the risk of death from lung cancer for the study group was much less than that from carcinoma of nasal sinuses, the mortality rate for lung cancer followed a similar pattern. The rate was highest among workers hired before 1920, with an average that was six to 11 times the national average. A continuous decline in the mortality rate was then observed. Thus, for those who started work in 1920-24 and 1925-29, the risk was reduced to 5.2 and 2.5 times the national average, respectively. A nonsignificant increase in the mortality rate of those employed in the period 1930-44 was observed; it was only 1.5 times the national average.

Barton (Ref. 129) discussed the role played by some nickel compounds in the causation of cancer of the nose and paranasal sinuses. He also recommended "a strict monitoring of the employees by regular examination and plasma nickel testing."

The presence of nickel in welding fumes from stainless steels, Inconel alloys, other specialty metals, and some welding rods is well documented. No incidents of the typical nickel worker's cancer of the nasal sinus have been documented as occurring in welders. No documented instances of lung cancer in welders attributable to nickel are recorded. Therefore, because nickel, a proven carcinogen, is found in some welding fumes, this does not mean that the nickel present causes the fume to be carcinogenic, or that the fume is carcinogenic. Pantucek's discussion of fluoride and Barton's discussion of nickel versus chromium indicate the possible decrease in biological activity caused by complex interactions in a plasma that is rapidly cooled to ambient temperatures.

Although chromium compounds have long been used in industry, not until the early 1950's were chromium's carcinogenic properties generally accepted. It is now widely agreed that only hexavalent chromium poses a carcinogenic danger; the trivalent element is excluded from the suspected list. Epidemiological studies showed that the target organ for the Cr VI carcinogenic effect is the respiratory tract, with a latent period of 10 to 30 years (Ref. 34). Stern points out that not all hexavalent chromium compounds are carcinogenic. Those that are water soluble are not carcinogenic, while those that are carcinogenic exhibit little or no solubility. He likens the carcinogenic effect of chromium VI to that of asbestos i.e., it implies a foreign body effect that depends upon the morphological and surface properties of the substance in its solid form.

Cancer of the Bladder

In a survey of 10 occupational groups, four seemed especially susceptible to bladder cancer. These are structural metal workers, boilermakers, plumbers, and welders (Ref. 130). This conclusion was based on the

*microgram percent = micrograms per 100 mL of fluid.

work of Milham (Ref. 50), who analyzed the mortality patterns associated with exposure to metals, and who maintained, "Welders and flame cutters show a lung cancer excess and excess mortality from other lung-related diseases. Chronic bronchitis with emphysema, other pneumoconiosis, and chronic interstitial pneumonia all show PMR (proportionate mortality ratio) increases in this group, as does urinary bladder cancer." Similar findings were reported by Howe et al. (Ref. 131) in a case-control study that involved 532 cases of bladder cancer in three Canadian provinces. The danger in using PMR studies alone has already been pointed out. The PMR can only indicate, in grossest form, areas for further research. To use a PMR as conclusive evidence alone to allocate cause of death is foolish.

Absorption, Distribution, Biotransformation, and Excretion of Fumes and Other Components

Fluorides can amount to 10 to 20 percent of the total particulates in the fumes from basic electrodes. It is well known that fluorides are readily absorbed from lungs as well as from the gastrointestinal tract. They also rapidly appear in the urine and the input/output balance is quickly established. Chronic exposure to fluorides leads to their deposition in bone tissues; then they may be excreted in urine in high concentrations for a long time after the cessation of exposure.

This led Pantucek (Ref. 51) to study welders using basic electrodes (type E234 B26 ISO). He found a close correlation between the concentration of fluorides in air and fluoride in the urine of welders (urine collected four hours after the start of exposure). The concentration of fluorides in the breathing zone was found to be 1.15 ± 0.58 mg F/m^3. The time-weighted average was 0.5 mg F/m^3. The author ascribed the lack of adverse effects, despite high fluoride intake, to the possible formation of a complex between iron and fluoride. Pantucek's hypothesis is that the iron content of welding fumes is high enough to fix all the fluoride present and thus lower its biological activity.

Epidemiologic Studies

In July 1976, NIOSH conducted a study of the adverse effects of prolonged exposure to cadmium, copper, zinc, and silver fumes, among other contaminants, present in bending and brazing operations (Ref. 132). Thirty-eight persons participated in this study: 20 braziers with services of 0.25 to 20.0 years (mean 7.9 years) who worked with a cadmium solder (12 females and eight males with a mean age of 37.9 [range 25-60] years) and an 18-person control cohort matched as closely as possible for age, sex, and smoking history (eight females and 10 males with an average age of 36.3 [range 23-53] years). The brazing materials used contained 0.5 and 35 percent silver. Of the cohorts, 14 exposed workers (70 percent) and 10 controls (55 percent) were smokers with a mean pack/year consumption of 27.8 and 22.5, respectively (one pack/year = one package of cigarettes per day for one year).

The following tests were performed on these cohorts:

(1) Medical history
(2) Physical examination
(3) Pulmonary function studies
(4) Blood pressure measurements
(5) Blood analysis
  (a) Complete blood count
  (b) Blood urea nitrogen
  (c) Blood cadmium level
  (d) Serum creatinine
  (e) Serum alpha-1-antitrypsin
(6) Urine analysis
  (a) Urine beta-2-microglobulin
  (b) Urine cadmium level
  (c) Urine uric acid concentrations
  (d) Urine creatinine concentrations
(7) Hair cadmium levels

During this study the average cadmium concentration in air averaged 0.074 mg/m^3. Previous determinations showed an exposure range of from 0.005 to 0.366 mg/m^3.

A proprietary tooling oil, as well as 1, 1, 1-trichloroethane (TCE), was found. Concentrations of the tooling oil ranged from 0-55 ppm, of TCE from 0-17 ppm (average 4 ppm). Silver was found in only small amounts on 9 of the 62 filter samples. The highest concentration was 0.015 mg/m^3. Copper ranged up to 0.089 mg/m^3 with average concentrations of 0.026; 0.017, 0.006, and 0.013 mg/m^3, depending upon the day's samples. Zinc had a high level of 0.281 mg/m^3, but the averages were 0.033, 0.013, 0.022 mg/m^3. Fluoride concentrations ranged up to 0.419 mg/m^3 with an average of 0.090 mg/m^3.

Responses to medical questions are shown in the following table, which indicates that chest tightness and headache are the only two symptoms that occurred more frequently in exposed workers.

The cutting oil was found to be an irritant, but probably not a sensitizing agent. A careful analysis of the laboratory data revealed no significant differences ($p < .05$) between exposed and control groups. The only variable shown to be different was the urine creatinine, but the significance of this is unknown. Three workers studied showed impaired or abnormal biomedical functions. The moderate-to-severe pulmonary disease indicated may have been due to cadmium exposure over a long period.

Exposure to copper, zinc, silver, and fluoride at the concentrations measured in this study were rated as nontoxic. Cadmium exposure, although sporadic, should be limited as much as possible.

The reason for including this study in a monograph on welders is that health specialists often confuse welding, brazing, and flame cutting. This study indicates that cadmium could possibly present a danger to brazers.

Epidemiological studies discussed later are grouped according to the disease in question. These diseases were: lymphosarcoma, pulmonary diseases, skin injuries, and urinary bladder cancer.

In an extensive study on the environmental factors in the epidemiology of lymphosarcoma, Goldsmith an Guidotti (Ref. 133) stated that "among the exposure which may be relevant to high rates of lymphosarcom in electrical workers are those of soldering fumes...." Th is based on proportional mortality ratio (PMR) of 14 in electricians. Welders are not mentioned among thos occupations studied.

Comparison of symptoms by history

Symptoms	Exposed % positive	Control % positive
Cough	60%	50%
Sputum production	50%	39%
Chest tightness	45%*	17% $P<.05$
Wheezing	30%	22%
Shortness of breath	40%	44%
Frequency of urination, L/night	10%	17%
Protein, sugar, or blood in urine	5% (sugar)	0%
Kidney or bladder problems	35%	28%
High blood pressure in past	10%	22%
Chest pain	20%	17%
Heart attacks	0%	0%
Abnormal blood fats	0%	0%
Miscarriages (female)	42%	0%
Difficulties with pregnancy (female)	8%	0%
Deformities in living children	8%	0%
Difficulties fathering children	0%	0%
Headache	75%*	28% $P<.01$
Dizziness	60%	39%
Trouble sleeping	25%	17%
Tremor	5%	6%
Abnormal sense of smell	5%	17%
Weight loss	10%	11%
Anemia	5%	6%
Fatigue	30%	22%
Loss of appetite	5%	6%
Skin rash	20%	6%
Hepatitis or liver disease	5%	0%
Frequent colds	5%	6%
Family members who have died or who have emphysema	35%	17%

Test	Exposed	Control
Pulmonary function test FEV_1/FVC	0.809 ± 0.076	0.838 ± 0.070
Blood pressure (abnormal $>$ 140/90)	35%	22%
Blood analysis		
White blood cells (1000/mL)	6.5 ± 1.30	6.8 ± 2.0
Red blood cells (millions/mL)	4.8 ± 0.45	4.71 ± 0.37
Hemoglobin	14.4 ± 1.80	14.3 ± 1.1
Hematocrit %	43.1 ± 4.8	42.6 ± 3.1
Blood urea nitrogen (mg/DL)	15.1 ± 4.2	16.0 ± 3.5
Serum creatinine (mg/DL)	1.2 ± 0.2	0.8 ± 0.1
Serum alpha-1-antitrypsin	268.5 ± 64.7	286.0 ± 89.4
Urine analysis		
Urine cadmium levels (μg/L)	10.55 ± 6.34	8.28 ± 4.82
Urine uric acid (mg/DL)	48.6 ± 24.4	38.0 ± 21.4
Urine creatirine (mg/DL)	124.9 ± 72.2	94.0 ± 59.2
Hair cadmium level (μg Cd/g hair)	2.62 ± 2.2	1.83 ± 1.90

***Symtoms showing suggestive differences at or approaching statistical significance.**

Pulmonary Diseases

In an article published in 1975, Pikulhskaya and Gulhko (Ref. 134) examined 473 electric welders for respiratory diseases. Of these, 96 showed signs of bronchial insufficiency. Among other complications, the authors found chronic bronchitis, obstructive emphysema, and pulmonary lesions.

Skin Injuries

An epidemiologic study conducted by Buncher et al. in 1977 (Ref. 52) revealed that skin injuries and the occurrence of cough in the morning in welders are the most significant findings. This study focused on the occurrence of cutaneous, ocular, and respiratory diseases in a group of 77 welders as compared to two other groups: (1) a control group of 58 employees who worked in different areas than the welders, and (2) a group of 75 workers who are not welders but were in the vicinity of the welding operations (exposed). Skin injuries in the form of acute erythema and cutaneous scars were common in welders, while the observed forced expiratory volume in one second (FEV_1) was lower among welders as well as among the "exposed" group.

Urinary Bladder Cancer

In 1976 Milham (Ref. 50) studied cancer mortality patterns associated with exposure to metals. He analyzed the Washington State death records by the proportionate mortality ratio (PMR) method. The occupations selected for study were: boilermakers, copper smelter workers, machinists, metal molders, plumbers, structural metal workers, sheet metal workers, aluminum millworkers, tool and die makers, and welders. His findings in the case of welders and flame cutters are cited in their entirety in the following table.

Proportionate Mortality Ratios (PMRs) – White male deaths in Washington State, 1950-1971, welders and flame cutters

Cause of death	Deaths			
	I.C.D. Number*	Obs.	Exp.	PMR**
All malignant neoplasms	140-204	257	246	104
Respiratory cancer	160-165	92	77	119
Cancer of trachea, bronchus, and lung	162.1	67	49	<u>137</u>
Cancer of urinary bladder	181	12	7	162
Chronic bronchitis with emphysema	502.0	7	1	<u>537</u>

*I.C.D. - International Classification of Diseases, 7th Revision.
**PMR underlined if 0.05.

The author suspected that urinary bladder cancer is due to the presence in the urine of carcinogens in the form of metals or metallic urinary excretion products.

The incidence of bladder cancer and other selected cancers in welders as compared to other metal workers is shown in the following table.

Selected cancer in metal workers — White male deaths in Washington State, 1950-1971

Occupation	Deaths (PMR) Cancer of tongue 141*	Cancer of pancreas 157*	Respiratory cancer 160-165*	Cancer of urinary bladder 181*	Malignant lymphoma 200-202* (excl. 201)
Boilermakers	87	93	150	144	100
Smelter workers	0	88	164	20	75
Machinists	180	96	105	92	95
Metal molders	0	46	135	100	100
Plumbers	152	75	130	137	146
Structural metal workers	295	99	108	203	100
Sheet metal workers	0	132	131	113	100
Tool and diemakers	0	52	155	0	100
Aluminum millworkers	0	204	128	95	250
Welders and flame cutters	58	63	119	162	111
Total	119	93	123	118	115

*I.C.D. - International Classification of Diseases, 7th Revision.

Human Fatalities

During welding, fatal injuries could be caused by machinery and heavy equipment as well as by the less spectacular but quite as real danger posed by welding fumes.

Deaths due to physical trauma have not been reported during the period 1978-79. However, fatalities attributed by the authors to exposure to welding fumes are summarized in the following table.

Exposure	Duration of exp.	Age, sex, and other characteristics	Cause of death	Reference
Cadmium fumes	Not mentioned	Male	Massive pulmonary edema, pulmonary thrombosis, kidney damage	Wolf (Ref. 44)
Cobalt aerosols	5 years	31-year old white female	Pulmonary fibrosis and cor pulmonale	Forrest et al. (Ref. 53)
"Black welding flux was introduced under the nail"	13 years	59-year old white male	Pulmonary metastasis	Raynor et al. (Ref. 48)
Zinc fumes	22 years (previously a coal miner)	52-year old male	Brain tumor and aphasia	Jaremin (Ref. 17)

Chapter 3

Toxicologic Investigation in Animals

Experiments on laboratory animals represent one of the main sources of information on the toxicity of various chemicals. To simulate the occupational exposure of workers to hazardous substances, animals are usually exposed to these compounds via inhalation. In view of this, and in order to study the effect of welding fumes, it is logical to assume that experimental animals should be placed in certain chambers for nasal exposure to welding fumes that should be generated "on the spot." Unfortunately, this is not the case because of the complex and variable composition of welding fumes. The majority of authors singled out one or more of the components expected to be present in welding fumes and tested the effect on experimental animals. This approach is objectionable for the following reasons:

(1) It falsely assumes that the biological effects of the compound will be the same if it is given alone or in conjunction with a number of other substances. Chemicals given simultaneously to any biological system are known to exhibit potentiation, synergism, or antagonism, or simply an addition of their effects. A familiar example is the antagonism between lead and zinc, both of which are normal constituents of welding fumes.

(2) A given compound may exhibit different physical properties when present alone than when in combination with others. Physical properties such as solubility, electric charge, and particle size, to name a few, are important in determining the outcome of biological experiments.

(3) Chemical properties also may change if a given compound is co-administered with others; e.g., alkaline fumes of welding might cause eye irritation, but when in contact with another acidic fume, may be neutralized and cause no irritation to the eye. It is well known that the pH plays an important role in toxicological effects.

(4) Different salts cannot be used to predict the effects of closely related compounds; e.g., the toxicity of potassium cyanide is totally different from that of potassium ferrocyanide. By the same token, the effect of manganese chloride cannot be extrapolated to manganese oxide.

(5) As mentioned earlier, the route of exposure must simulate those routes that are prevalent in occupational situations. Thus, results of experiments in which iron oxide is injected intramuscularly cannot be used to predict siderosis in welders. Even tracheal instillation of a given compound is unreliable, since it might provoke different reactions than when given via inhalation. The concentration of the compound at the site of instillation may be so great that it induces a local lesion that initiates a totally different pathological picture of the lung than if the same compound is given through inhalation.

(6) Results of experiments in which there is total body exposure of the animals should be interpreted with caution. This is because animals exposed in this way are apt to have the aerosolized compound precipitated over the skin. Animals normally clean themselves by licking their skin often; thus, the exposure effect is intensified. If the compound is absorbed through the skin or the gastrointestinal tract, the toxicological profile from inhalation of that compound may be intensified, modified, or even completely changed.

In view of the above mentioned factors, the best experimental design to study effects of welding fumes is the one in which animals are exposed (head or nose only) to the total fumes generated on the spot. This approach provides the best approximation to human occupational exposure in which the composite effects of fumes, heat, radiation, and noise are present.

Methods for the controlled generation of welding fumes for a period of several hours, as well as methods for characterization of these fumes, are detailed by Hewitt et al. (Ref. 135).

Welding fumes are extremely complex materials whose composition will vary depending upon the metal being welded, the method of welding, and a host of other factors. For this reason, only three articles have appeared since 1978 in which the effects of welding fumes on experimental animals were studied. The vast majority of the experimental studies used one or more materials that are chemically pure and can be easily identified. The discussion of these biological investigations will, therefore, be divided into two major parts: (1) studies using welding fumes, and (2) studies performed on the various components of the fumes.

Experiments Using Welding Fumes

A system for the generation of welding fumes to study their effects on experimental animals was mentioned by Hewitt and Hicks in 1972 (Ref. 136). Although this article is almost eight years old, it is cited because, with some modification, the system described could be ideal for animal studies.

This system is composed of a welding cowl in which fumes are produced by regular welding procedures. Fumes thus generated are drawn through a centrally placed, perforated tube into the exposure chamber (see Fig. 3.1). Fumes then were exhausted by a pump through a device to collect the fumes on a filtering pad. The concentration of fumes was then calculated from the flow rate and the weight of collected fumes.

The merit of this system is that fumes are freshly generated; thus, their composition and temperature simulate closely those under actual welding conditions. There are three main drawbacks:

(1) As discussed previously, animals should be exposed to fumes, nose only.

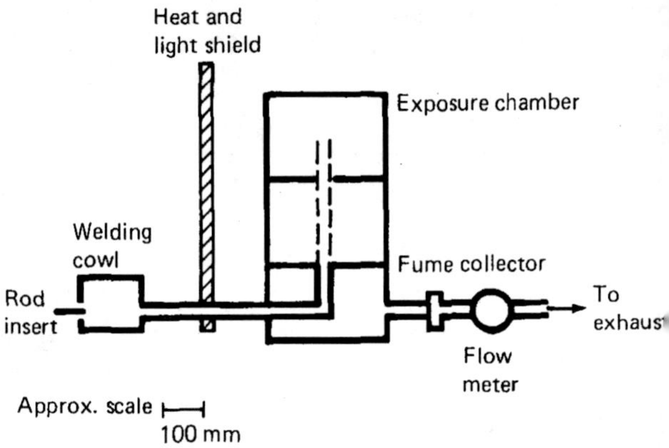

Fig. 3.1

(2) There should be some sampling holes at dif ferent sites in the chamber so that various sample could be withdrawn to determine whether or not there is an equal distribution of the fumes.

(3) The calculation of the fume concentration, a shown by the authors, ignores the amount already de posited in the animal's lung, on its body, or in th chamber, which may be significant.

These drawbacks led the authors to design anothe exposure chamber in which the above-mentioned point were taken into consideration.

Male albino rats were used to study the effects o welding fumes produced by striking standard rutile iro welding rods (Type C18) on a steel workpiece. Animal were then exposed for various periods of time. The firs two groups (one rat per group) were exposed for 1 and 30 minutes, respectively. Twelve rats in the thir group were exposed to fumes for 235 minutes. Th fourth group consisted of 5 rats and was exposed t fumes for nine days, according to the following schedule

Day	Duration of exposure (min)
1	33
2	31
3	30
4	30
5	32
6, 7	No exposure
8	30
9	29
10	26
11	32

The composition of the fumes were as follows:

Group no.	Exposure duration (min)	No. of rats exposed	Concentration of elements in the fume (mg/m^3)				
			Sb	Co	Cr	Fe	Zn
1	16	1	0.72	0.28	0.02	392	0.01
2	30	1	0.46	0.18	0.01	250	0.01
3	235	12	0.21	0.09	0.01	143	0.01
4	See text	5	0.95	0.38	0.03	514	0.02

No physiological response was observed in groups one or two. On the other hand, animals in groups three and four showed signs of nasal irritation and coughing at the end of exposure. However, upon cessation of exposure, the animals recovered quickly. Except for the lungs, histopathological examination revealed no abnormalities. Lungs showed an abundance of macrophages and also a slight thickening of the epithelial cells.

In an attempt to study the effect on lungs, Senczuk and Mlynarczyk (Ref. 137) administered a suspension of welding dust in sodium chloride solution intratracheally to rats. An increase in the amino acid hydroxyproline content of the lungs was observed, while the content of free amino acids in the plasma was remarkably reduced. It was suggested that the increase in the amino acid contents of the lung is indicative of fibrosis of pulmonary tissues. This experiment does not add to our knowledge of welding fume toxicity, since instillation of a suspension of welding fume into the trachea is not the same as inhaling the fumes.

The other study simulated the actual occupational atmosphere in that the animals (rats and guinea pigs) were allowed to inhale welding fumes for up to five hours. Two different types of welding fumes were utilized: those from the flux cored electrode (FCAW) and the metal inert gas (GMAW) welding processes. The concentration range of the fumes was 900 to 1200 mg/m^3.* Unlike the first study, this investigation was designed to study the rate of clearance of the fumes from lungs. The half-life (the time in which 50 percent of the material deposited is cleared) of alveolar clearance was 19 to 35 days and 70 to 200 days for FCAW and GMAW fume components, respectively.

Doi (Ref. 138) examined the toxic effects of welding fumes that contain antimony as compared with those that do not. It was interesting to note that the amount of fumes generated from a welding rod containing antimony was lower than those generated from antimony-free rods. However, guinea pigs exposed to the fumes that contained Sb_2O_3 and Sb_2O_5 showed no significant difference from those that were exposed to Sb-free fumes. The highest concentration of antimony was found in the lungs one hour after exposure. The antimony level gradually decreased, reaching 10 percent of its concentration after one week, and only a trace could be detected after three weeks.

Effect of Fumes on the Skin

Welding fumes generated by the manual metal arc (SMAW) process or by the automatically controlled metal inert gas (GMAW) process were studied for their ability to induce experimental hypersensitivity in female guinea pigs (Ref. 139). Analysis of the elemental constituents of these fumes showed the following composition:

Fume particles % W/W	SMAW	GMAW
Iron	32	61
Cobalt	0.009	0.017
Nickel	Undetectable	6
Chromium	0.004	11
Silicon (as SiO_2)	18	Undetectable

Various methods were used to test for the induction of skin hypersensitivity. These methods included guinea pig maximization test, split adjurant technique, and other methods. It was found that GMAW material produced 23 unequivocally positive reactions in a total of 40 animals; SMAW material was less effective, with only 10 unequivocal reactions out of 40. Although the results of this study on experimental animals are well documented, their extrapolation to human risk is not an easy task. Primarily, this is because welders are exposed to these fumes mainly by inhalation, which means that various elements of welding fume deposits might be eluted into tissue fluids, or translocated to the drainage lymph nodes, and lead to hypersensitivity reactions. Also, hypersensitivity caused by prior industrial exposure to various metals might complicate the net result of exposure to metal fumes.

Experiments Using Individual Components of the Fumes

In this section, experiments using Be, Cd, Co, Fe, Mn, and Ni in experimental animal models are reported.

Beryllium

An *in vitro* study using rat liver parenchymal derived cell line showed that toxicity of beryllium compounds is related to the degree of their uptake by these cells (Ref. 140). Previous studies from the same laboratory (Ref. 141) showed that the soluble beryllium sulfate is more toxic to rats' livers than particulate beryllium phosphate.

Cadmium

Rats were used by Conklin et al. (Ref. 142) to study the effects of cadmium oxide. Animals were acutely exposed to 60 mg/L cadmium oxide aerosols for 30 minutes. Almost 50 percent of the rats died within three days from massive pulmonary edema. Tissue analysis showed a high concentration of cadmium in the liver. Animals that survived were sacrificed a year later and over 80 percent of them exhibited seminiferous trubule degeneration, as compared to 21 percent of unexposed rats. Only one animal showed adenocarcinoma of the lung as opposed to none in the control group.

The effects of cadmium on the airway cell response

*This range of fume concentration is not typical of ordinary welding conditions.

was studied *in vitro* by Gardner et al. (Ref. 143) and *in vivo* by Asvadi and Hayes (Ref. 144). In the first study, isolated hamster tracheal rings were exposed to various concentrations of cadmium chloride. A concentration as low as 0.006 mM (millimoles) of cadmium was found to reduce significantly the ciliary activity of the epithelium. This finding is of great importance as it shows that cadmium inhalation impairs the normal mucociliary function responsible for clearance of foreign particles from the respiratory tract.

The objective of the *in vivo* study was to investigate the effects of cadmium aerosol on the cell population of the lung. Rats given a single exposure to 0.005 M of cadmium chloride showed an increase in the polymorphnuclear leukocytes, alveolar macrophage, and lymphocytes as compared to the control group, which received physiologic saline only.

In 1977, Bouley and co-workers (Ref. 145) used rats and mice to study the effects of microparticles of cadmium on the respiratory system. Animals were exposed to a single 15 minute exposure to 10 mg/m^3 of cadmium oxide and then were observed for a 24 day period. About 4 mg of cadmium were retained by the lungs of the rats. Exposure of the rats led to an initial decrease followed by an increase in polymorphnuclear and lymphocytic cells in the alveoli. Challenge of animals with a test infection with Salmonella enteritidis led to a higher death rate in animals exposed to cadmium as compared to the control group. The authors ascribed cadmium toxicity to its cytotoxic and enzyme-impairment effects. These results are in agreement with those of Gardner et al. (Ref. 143) who found a higher mortality rate from viable streptococci in rats that are pre-exposed to cadmium chloride (80 to 1600 mg/m^3). The increase in mortality was concentration-dependent and ranged from 15 percent in the smallest concentration to 70 percent in the highest. The authors found a close correlation between the rate of clearance of streptococci from the lung following exposure to cadmium chloride and the observed mortality rate.

Acute exposure of rats to 0.1 percent cadmium chloride aerosol produced lesions in the lungs characterized by necrosis of type I cells within 48 hours of exposure (Ref. 146). Marked mitotic activity of type II cells soon followed. These two stages were accompanied by a tremendous increase in polymorphnuclear leukocytes and alveolar macrophages, respectively.

Acute exposure of rats to 19.5 mg/m^3 of cadmium oxide for 90 minutes resulted in a decrease in the number of macrophages in the lung (Ref. 147). Koshi et al. claimed that the toxicity of cadmium was enhanced by the fact that cadmium oxide is readily soluble in the biological fluids of the lung.

To determine the absorption rate and fate of cadmium, Boisset et al. (Ref. 148) exposed male Sprague-Dawley rats to 9.43 mg/m^3 of cadmium oxide fumes, 30 minutes per day for five consecutive days. About 70 percent of the particles in the fume were less than 0.5 μm in size as determined by electron photomicrography. The rats were then sacrificed at different times within a three-month exposure period. Results showed that about 12 percent of the inhaled cadmium was deposited in lungs, leading to some degree of permanent damage to pulmonary lobes. Absorbed cadmium led to a decrease in liver growth but not in kidney growth.

Pulmonary edema that developed in rats after exposure to aerosols of cadmium oxide (60 mg/L for 30 min) could be reduced by pretreatment with the radioprotective agent WR 2721 (S-2-amino propylamino) ethyl phosphorothioic acid hydrate (Ref. 149).

Hemodynamic Effects

Prigge et al. (Ref. 150) used male SPF-Wistar rats to study the effects of cadmium on the hemopoietic system. Cadmium chloride was administered either by inhalation (0.2 mg Cd/m^3, for 66 days) or orally by adding 25, 50, or 100 ppm Cd (as $CdCl_2$) to drinking water for 48 to 55 days. Cadmium given by inhalation resulted in no significant change from the control group insofar as hematocrit and hemoglobin percentages are concerned. However, all groups of animals that received cadmium orally showed an anemic response.

Prigge (Ref. 151) exposed pregnant and nonpregnant rats to cadmium aerosols ranging from 0.2 to 0.6 mg cadmium/m^3, for 21 days without cessation. Nonpregnant rats showed a polycythaemic response due to the stimulant effect of cadmium on erythropoiesis. Pregnant rats showed a less marked polycythaemia. No proteinurea could be detected either in pregnant or nonpregnant rats. To confirm these results, the author (as described in another publication [Ref. 152]) exposed female rats to 25 to 50 mg/m^3 of cadmium oxide aerosols for 90 days. Another group of rats was exposed to 100 mg/m^3 for 63 days. Cadmium was also given orally to another set of animals that received 25, 50 and 100 ppm in their drinking water for 90 days. Although both routes of administration resulted in comparable kidney cadmium levels, proteinurea was observed only after the oral administration. High cadmium levels in blood and liver were observed after oral administration. On the other hand, inhalation led to a marked dose-dependent weight increase of the lungs. Impairment of gas exchange was ascribed to lung damage. Serum iron was significantly reduced after oral administration of cadmium but not after inhalation.

Biochemical Effects

Excessive exposure to ozone or nitrogen dioxide could lead to pulmonary lesions that are characterized by an initial destruction of type I alveolar pneumocytes associated with the development of pulmonary edema. From a biochemical point of view, exposure to these substances leads to an increase in the activity of lung enzymes glucose-6-phosphate dehydrogenase, glutathioveperoxidase, and superoxide dismutase. These enzymes are known to prevent the membrane-damaging effect of lipid peroxidation. The increase in the activity of these enzymes constitutes a defense mechanism and explains tolerance observed with repeated exposure to these agents. Exposure induces not only biochemical changes but also alterations in pulmonary physiology parameters such as tidal volume and frequency of respiration.

Bus et al. (Ref. 153) found that such biochemical and physiologic changes in the lung could be also induced by cadmium. To study this effect, they exposed male rats to 0.5 percent aerosol of cadmium chloride for one hour. Acute effects observed one day after exposure included a significant decrease in tidal volume and a remarkable increase in respiratory rate. The lung-to-body-weight ratio was 192 percent as compared to the control group. A zero order of cadmium clearance was found with a half-life interval of 27.4 days. Biochemical changes, on the other hand, were manifest after 11 days from the start of exposure. These changes included an increase in the activity of glucose-6-phosphate dehydrogenase as well as superoxide dismutase enzymes.

Fetotoxicity

Pregnant rats were continuously exposed to 0.2, 0.4, and 0.6 mg/m^3 of cadmium for 21 days. Despite relatively effective placental barriers to cadmium, it was found that at 0.6 mg/m^3 fetal growth rates were reduced. The activity of alkaline phosphatase was affected in a manner that was dose-dependent. While that of the fetus increased gradually with the increase in dosage, maternal alkaline phosphatase gradually decreased (Ref. 151).

Cobalt

Cobalt is sometimes added to steel and various alloys to enhance certain characteristics of these metals. It follows that welders could be exposed to cobalt in the fumes generated when such alloys are welded. Hamilton and Hardy (Ref. 154) maintain that cobalt could produce damage to lungs if combined with other insults such as cigarette smoking. This led Wehner et al. (Ref. 155) to study the effects in hamsters of the chronic inhalation of cobalt oxide alone and in conjunction with cigarette smoke. Exposure to cobalt oxide resulted in pneumoconiosis, but the life span of the animals was not affected. No malignancy was observed. However, animals that were exposed to cobalt oxide and cigarette smoke showed an increase incidence of tumors and, surprisingly, increased life span.

In another study by Georgiadi and Elkind (Ref. 156), rats were exposed to 0.48 mg/m^3 of powdered metallic cobalt. This led to some dystrophic changes in the mucosa of the respiratory tract.

Biochemical Effects

Georgiadi (Ref. 157) examined the effect of chronic exposure to cobalt (0.48 mg cobalt dust/m^3) on the respiratory mucosal enzymes of rats. A two-month exposure decreased the activity of all the enzymes tested (succinate-, glutamate-, glycero-, phosphate-, 6-phosphate, glucose-6-phosphate,- lactate-, isocitrate-, and maleate-dehydrogenases, adenosine triphosphatase, alkaline phosphatase, and esterase). Further exposure for two more months caused an increase in the activity of all studied enzymes except 6-phosphogluconate-, maleate-, and isocitrate-dehydrogenases and esterase, which remained below normal. The changes in these enzymes for another group of animals exposed to a higher concentration (4.4 mg cobalt/m^3) for two months, differed according to the animal's gender. An additional two months of exposure resulted in an inhibition of these enzymes regardless of the sex of animals.

Iron

Guinea pigs were injected intratracheally with iron ore dust (98.1 percent ferric oxide, 0.66 percent silica) in the dose of 75 mg suspended in 1.5 mL of physiological saline. Macroscopic examination of their lungs showed patchy congestion that subsequently involved the entire lobe (Ref. 158).

The effect of iron oxide particles on the pulmonary macrophage, the main pulmonary defense system, has received the attention of several investigators at the Harvard School of Public Health. In 1977, Levens et al. (Ref. 159) exposed male mice for three hours to an iron oxide aerosol (Fe_2O_3) generated by the combustion of iron pentacarbonyl, $Fe(CO)_5$, vapors. The concentration of Fe_2O_3 in air ranged from 50 to 400 mg/m^3. To evaluate the effect of exposure on pulmonary alveolar macrophage, animals then were exposed to aerosols of radiolabeled *S. aureus*, one hour after the exposure to Fe_2O_3. It was found that exposure of animals to Fe_2O_3 decreased the respiratory minute volume as well as the ability of macrophages to defend the lung against bacterial invasion.

In another study by the same group reported by Kavet et al. (Ref. 160), Syrian golden hamsters were exposed to Fe_2O_3 aerosol for 3 hours, then the *in vitro* phagocytic rates of lavaged pulmonary macrophages were assayed. It was found that iron oxide stimulated the phagocytic rates and provoked a recruitment of mononuclear phagocytes into the pool of cells.

In the third article by the same investigators reported by Watson and Brain (Ref. 161), the ability of the tracheal and bronchial epithelium of male mice for the uptake of iron oxide was assessed. Animals were exposed to iron oxide (300 mg/m^3) aerosol for three hours. Iron uptake as shown by the electron microscope revealed that all types of epithelial cells, except mucous cells, are capable of pinocyting iron oxide, which is then converted to ferritin and hemosiderin.

Scanning electron microscopy was used by Quan and Golde (Ref. 162) to study the *in vitro* effect of carbon and iron particles on the morphology of human alveolar macrophages. The addition of iron particles to the culture medium resulted in an approximately tenfold increase in the membrane convulsions. The authors claim that these morphological changes reflect the functional and metabolic alterations resulting from exposure to such foreign particles.

The role of the lymphatic system in the clearance of certain metals from the lungs was studied in dogs that were exposed by intrabronchial administration (Ref. 163). It was found that lymphatic uptake of iron is an important channel of alveolar clearance, while the predominant clearance mechanism for lead is nonlymphatic.

Manganese

Following reports that manganese dust causes some neurological and sexual disturbances among manganese ore crushers, Singh et al. (Ref. 164) elected to study the early biochemical signs of toxicity of manganese dioxide in rats. The investigators used male albino rats to which 50 mg of manganese dust suspended in 1 mL saline was instilled intratracheally. Thirty days after the inoculation, the rats were exsanguinated and the lung tissue was examined to determine manganese content as well as enzymatic activity. No significant histopathological nor biochemical alterations were observed. However, there was a marginal increase in the activity of SDH and NADH-cytochrome reductase. These changes suggest a slight enhancement in energy metabolism, presumably for phagocytosis. The authors also speculated that the increased activity of phosphoglucomutase and the slight inhibition of aldolase noticed in manganese-exposed lungs are not likely to affect glycolysis.

Chronic toxic effects of manganese administered by inhalation were studied by Griffin and Coulston (Ref. 165), who exposed rats and rhesus monkeys to particulate manganese at 100 mg/m^3, 23 hrs/day for 8 weeks (rats) or one year (monkeys). Particulate manganese was generated by the combustion of methyl cyclopentadienyl manganese tricarbonyl. There was a small increase in urinary excretion of manganese as well as increased concentration of the metal in various tissues, especially the lungs. No sign of untoward effect was observed, nor could any histological change be detected. Similar results were achieved by Ulrich et al. (Ref. 166) who exposed rats and monkeys (*Saimiri sciureus*) of both sexes to concentrations of Mn_3O_4, ranging from 11.6 to 1,152 mg/m^3 (expressed in terms of elemental manganese). Except for a slight increase in the hemoglobin and mean corpuscular hemoglobin concentration, no ill effects were noticed.

Similar results were discovered by Bergstrom and Rylander (Ref. 167), who found that an increase in the number of macrophages and leukocytes appeared one day after a 24-hour exposure of guinea pigs to manganese dioxide.

Biochemical Effects

To study the biochemical effects of manganese oxide, a group of Japanese scientists exposed rhesus monkeys to manganese dioxide dust at concentrations of 0.7 and 3.0 mg manganese/m^3, 22 hours/day, over 10 consecutive months. Two out of three monkeys exposed to the higher concentrations of manganese showed some neurologic manifestations, such as mild tremors of the fingers, loss of dexterity in the movement of the upper limbs, and a loss in pinching force. These signs developed three to four months after the start of exposure. Although they did not intensify, they lasted for the entire period of exposure. Biochemical changes included an increase in serum glutamic oxaloacetic transaminase (SGOT) and serum glutamic pyruvic transaminase (SGPT). A marked increase in monoamine oxidase (MAO) activity was noticed in the liver, adrenal gland, cerebellum, and mesencephalon. Adenosine deaminase activity was increased in the serum as well as in various parts of the central nervous system (cerebral white and gray matter cerebellum, diencephalon, mesencephalon, pons, medulla oblongata, and basal ganglia).

Manganese levels were also increased in various tissues especially the lungs, lymph nodes, and the cerebral basal nuclei. Central nervous system manifestations of toxicity were ascribed to the high concentrations of manganese in basal nuclei.

Nickel

The effects of different metals on the respiratory system of various animal species have been studied by various authors. In a study published in 1978, Camner et al. (Ref. 168) exposed two groups of rabbits to two different concentrations (0.5 and 2.0 mg/m^3) of metallic nickel dust for six hours a day, five days a week, for four weeks. In both groups there were dose-related changes in the lung weight and density as well as in the phagocytic activity. Changes were detected in the size distribution, and ultrastructure of alveolar macrophages of exposed animals. Some of the features of lung change due to nickel were likened to the alveoli lipoproteinosis that occur in rats exposed to silica, and to the pulmonary alveolar proteinosis encountered in man.

The cytotoxic effects of nickel on ciliated epithelium were studied by Adalis et al. (Ref. 169) using isolated hamster tracheal rings. As in the case of cadmium, nickel caused a decrease in ciliary activity at concentrations as low as 0.011 mM. It was concluded that by decreasing the ciliary movements, and hence the normal defense mechanism, nickel could increase the subjects' susceptibility to respiratory infections.

Hemodynamic Effects

Erythrocytosis, an increase in the total red cell mass was induced by five nickel compounds when administered intrarenally to rats (Ref. 170). Results showed that nickel subsulfide (Ni_3S_2) was the most potent in inducing erythocytosis, followed by crystalline nickel monosulfide (NiS), nickel subselenide (Ni_3Se_2), and metallic nickel dust (Ni), in order of responsiveness. However, since the route of administration of these compounds to animals is not encountered in industrial exposure, the implication of this study for human beings cannot be easily defined. In another communication, Hopfer and Sunderman

(Ref. 171) found that erythrocytosis induced in rats by the intrarenal injection of Ni_3S_2 is inhibited by simultaneous exposure to manganese dust.

Biochemical Effects

Male albino rats were used to investigate the biochemical effects of nickel on liver, kidney, testis, and myocardium (Ref. 172). In this study by Mathur et al., animals were injected intraperitoneally with 3 mg nickel/kg, as $NiSO_4 \cdot 6H_2O$ daily for 90 days. Five rats were sacrificed at different intervals (at 7, 15, 30, 60, and 90 days). No biochemical changes were noticed in animals sacrificed at seven and 15 days. Rats sacrificed on day 30 showed a significant increase in the activity of adenosine triphosphatase in the kidney and testis as well as in the kidney's acid phosphatase. There was a reduction in the activity of the phosphorylase enzyme in the heart. Continued exposure of animals to nickel for 60 days resulted in a significant decrease in adenosine triphosphatase activity in the kidney, but the increase in that of the testis shown before (30 days) was sustained. There was also a significant decrease in the activity of succinic dehydrogenase enzyme in the liver and kidney, as well as in that of acid phosphatase in the testis. Significant changes were found in the activity of all enzymes studied in the animal that received nickel for 90 days. Two enzymes showed increased activity, viz. adenosine triphosphatase of the testis and acid phosphatase of the liver. On the other hand, a significant decrease in the activity of the following enzymes was noticed: liver and kidney's succinic dehydrogenase; kidney's adenosine thiophosphatase; testis' acid phosphatase, as well as phosphaylase enzyme in the myocardium. Because nickel toxicity was manifested only after 60 and 90 days of the administration of that metal in such a small dose, the authors speculated a cumulative effect in these vital organs. At that dosage level, testis seemed to be the least sensitive organ studied.

Exposure to metals is known to affect not only the diffuse mechanisms of lungs, but even the primary humoral immune system. The effect of cadmium, nickel, and chromium on the humoral immune system of mice was studied by Graham et al. (Ref. 173). Results of that study showed that intramuscular injection of 3.9 mg/kg of nickel (as $NiSO_4$), or 9.25 mg/kg (as $NiCl_2$), caused a supression of the immune system.

There is growing evidence that exposure of the lungs to various environmental pollutants could result in a serious derangement of the defense mechanisms. This ultimately leads to an increase in the susceptibility of exposed animals or humans to infectious micro-organisms.

There are four major lines of defense against particulates that gain access to the respiratory system: (1) The so called "microciliary escalator" helps in removing particles deposited in the upper airways by moving the mucus in which these particles are imbedded upward toward the throat. Although it protects the lungs, this defense mechanism may not be entirely effective to protect the whole body, since these foreign particles might end up in the gastrointestinal tract where they undergo various degrees of elution and absorption. (2) Alveolar macrophages play a major role in clearing the deep lung tissues of deposited particles. (3) Direct intercellular penetration. (4) Solubilization or leaching.

It has been demonstrated that metals can exert toxic effects by hampering the function of respiratory tract cilia, thus leading to the accumulation of hazardous amounts in lungs. *In vitro* studies using hamster tracheal rings showed that cadmium and nickel exert this effect. Other metals such as chromium, manganese, and vanadium can also affect the alveolar macrophages in such a way that their phagocytic and enzymatic activity is decreased and their number and viability are reduced. Detailed discussion of the effects of these metals on the defense mechanisms of the lung is, therefore, warranted.

The toxic effect of metals and metal oxide particles in the welding environment can be modified by the presence of acidic or alkaline fumes. The reverse is also true. For instance, Charles et al. (Ref. 174) studied the effect of cadmium, cobalt, nickel, iron, manganese, and zinc cations on the absorption rate of sulfates in the rat lung. They found that manganese ions had no effect on the sulfate absorption rate, whereas cobalt, cadmium, and nickel enhanced it at a concentration of 0.1 mM/lung. Iron and zinc cations were the least effective, reaching maximal effects only at 10.0 mM/lung. Heavy metals are known to exert an effect on absorbing or excreting membranes. Local effects on the lung could possibly include the release of histamine, which might be responsible for the precipitation of asthmatic attacks.

Studies on Carcinogenicity of Metals

A group of Japanese scientists studied lung carcinogenicity of several kinds of dust, including copper ore and arsenic trioxide, administered by intratracheal instillation to male Wistar-King rats (Ref. 175). In this study 2.5 mg of copper ore, and 0.26 mg of arsenic trioxide were instilled intratracheally once a week for 15 weeks. No squamous cell carcinoma of the lung was found. However, the group of 10 rats that received copper ore showed one adenoma; those that were treated with arsenic trioxide (eight rats) also showed one adenoma. An important finding, however, is that arsenic trioxide, when administered with benzo (a) pyrene, seems to act as a co-carcinogen in inducing lung cancer in experimental animals. Benzo (a) pyrene is a constituent of coal tar.

The hypothesis that nickel could induce cancer was tested by Saknyn and Blokhin (Ref. 176) in white rats. Nickel monoxide was injected intraperitoneally, or given by inhalation or intratracheal instillation. It was found that intraperitoneal injection of nickel oxide induced sarcomas at the site of injection in six out of 39 rats, six to 15 months after the injection. Only one out of five rats showed squamous cell lung carcinoma without keratinization, following inhalation of nickel oxide; and one out of 26 rats showed the same type of carcinoma after intratracheal instillation. The authors conclude that dust at nickel production plants is blastomogeni Extending these ideas to welding fume is dangerou although nickel workers are apparently at high risk.

Chapter 4

In Vitro Studies

According to a current population survey in the U.S., there are an estimated 713,000 welders and flame cutters who are actively employed, and 52,000 who are unemployed (unpublished report of BLS, 1979). These welders are exposed in varying degrees to welding fumes containing, among other substances, nickel and chromium, which the Registry of Toxic Effects of Chemical Substances (Ref. 177) lists as suspected carcinogens in experimental animals. It has been suggested that about 60 to 90 percent of all cancer cases in humans may be related to environmental factors (such as sunlight and smog) and life-style factors (such as smoking and drinking) acting alone or in conjunction with genetic or other factors (Refs. 178-180). Therefore, much human cancer may be preventable if the causative agent(s) can be identified and human exposure to them minimized or eliminated.

A unique feature of cancer is that it usually develops in humans after a latent period of five to 40 years. This means that any increase in cancer victims due to various substances utilized at the present time is not yet observable.

Therefore, the best available scientific method for assessing cancer risk in welders is the analysis of data from human epidemiological studies, as well as the interpretation of results from adequately designed and conducted animal studies. The only drawback to these studies is the long time period required for completion, often several years.

In view of the need for results, several *in vitro* screening techniques have been developed to at least "point a finger" at potential carcinogens. These techniques are based on the assumption that cancer can be related to genetic alterations and, therefore, detection of such genetic changes might indicate the carcinogenic potential of various compounds. Although these tests use a variety of bacteria, yeast, and mammalian cells, the most widely used appears to be the Ames test, which depends upon the use of several specially constructed strains of *Salmonella* bacteria. Other *in vitro* tests include:

(1) Mutagenesis in bacteria, *Drosophilia melanogaster,* or in mammalian somatic cell culture

(2) DNA damage and repair

(3) Neoplastic transformation of mammalian cells

(4) Chromosomal damage

(5) Dominant lethal test

It should be noted that it would be technically unjustifiable to label a chemical as a carcinogen solely on the basis of these tests. For instance, it is well known that there are some mutagens that are definitely not carcinogens. It is estimated that false positive results occur in 2 to 10 percent of compounds studied for their mutagenicity (Refs. 181 and 182). The importance of these tests, therefore, is to raise a "red flag," especially when they are combined with positive findings in experimental animal models. With this in mind, we can examine the results of mutagenicity studies performed with welding fumes. Unfortunately, the only technical proof that a substance is a human carcinogen is the induction of cancer in humans by the substance in question.

Chromium compounds were studied for mutagenic activity by Bigaliev et al. (Ref. 183) using embryonal fibroblast cell culture and human lymphocytes. Potassium dichromate was found to be mutagenic in embryonal fibroblast. Other chromium compounds entering the human organism by inhalation were found to cause chromosomal aberrations.

Using the *Salmonella* histidine revertant test of Ames, Stern (Ref. 33) studied the mutagenic effect of water-soluble portions of stainless steel and mild steel welding fumes. He found that the latter is not mutagenic while stainless steel fume extract is mutagenic. The mutagenic activity of the fumes from flux cored arc welding (FCAW) and shielded metal arc welding (SMAW) with coated electrodes could be attributed, according to Stern, to the soluble hexavalent chromium in the fumes. He also attributed the revertant activity of fume from argon shielded arc welding to possibly both hexavalent chromium and nickel. These results could raise a false alarm, since only certain forms of hexavalent chromium are known to be carcinogenic in humans. This is a classic example of the extreme caution required when applying the findings of the Ames test to man. In Stern's laboratory tests, nickel was found to be nonmutagenic.

Nickel is a recognized carcinogen with a threshold limit value, for minimal health effects, of 100 micrograms per cubic meter. This limit value is high compared to the proposed chromium (VI) value of 1 microgram per cubic meter. Therefore, nickel does not presently pose a problem unless further research shows that a downward revision of the limit value for nickel should occur. If that downward revision does occur to a limit value of less than 1 microgram per cubic meter, then nickel will have been shown to be a more potent carcin gen than chromium (VI).

The same experimental design was used by Maxild al. (Ref. 54) to investigate the possible mutagenic effec of the fumes produced by various welding methods the *Salmonella* microsome. Fumes from two types welding, viz. flux cored arc welding (FCAW) a shielded metal arc welding (SMAW), as applied to tw different classes of materials – mild steel (ms) a stainless steel (ss) – were used in this study. Do response relationships have been studied on the extrac of fume generated from the following combinations materials, methods, and electrodes:

(1) FCAW/ss: A rutile type electrode with lo carbon weld metal of type 18/10, for use with 18/ type austenitic steel.

(2) FCAW/ss: A rutile type electrode with lo carbon weld metal of type 18.5/12.5 + 3% Mo, for u with acid-proof steel.

(3) SMAW/ss: A low carbon wire for welding 18/8 austenitic stainless chrome-nickel steel (argon ar

(4) SMAW/ss: A low carbon wire of compositi 18/8 + 2.5% Mo, for welding high corrosion-resista (acid-proof) steel (argon arc).

(5) FCAW/ms: A larger diameter, high produ tivity, rutile covered electrode typical for use in hea construction.

(6) SMAW/ms: A powder filled wire of rutile typ for use on ms with argon CO_2 mixture.

Fumes obtained by using short arc (19V) were al included to compare it with those from SMAW. T elemental composition of fumes from different ope tions were analyzed by atomic absorption and flar emission spectroscopy, ion-specific electrodes and prot induced x-ray fluorescence. Results, represented as p cent of weight, were as follows:

Element	Mild steel FCAW Acid-sol.	Mild steel SMAW Acid-sol.	Stainless steel FCAW Acid-sol.	Stainless steel FCAW Water-sol.	Stainless steel SMAW Acid-sol.	Stainless steel SMAW Water-sol.	Stainless steel FCAW (short arc) Acid-sol.	Stainless steel FCAW (short arc) Water-sol.
Al	0.25							
Na	0.56				0.068	0.032	0.088	0.027
Mg	0.30	0.45			0.013	0.004	0.16	0.12
F	0.12	1.0	10	9	0.07	0.05	0.21	0.21
Si	10.4	5.9	8.8		1.0		3.2	
K	6.6	1.7	15.2	14	0.043	0.027	0.046	0.022
Ca	0.50	0.76			0.15	0.046	0.18	0.006
Ti	3.5	3.2	0.91		0.05		0.05	
Cr	0.03		3.6	3.3	13.8	0.0097	9.8	0.041
Mn	5.2	15.1	2.3	0.2	10.4	0.24	7.0	0.56
Fe	36.0	36.6	5.0		36.9	0.002	37.7	0.008
Ni			0.50	0.05	4.1	0.079	6.5	0.25
Cu	0.12	0.22	0.33		0.11		0.13	
Zn	0.015	0.12	0.16		0.42	0.004	0.60	0.018
As	0.05	0.008						
Zr	0.04	0.13	0.01					
Nb	0.02		0.01					
Mo	0.05		0.77		0.36	0.02	2.3	0.43
Ag	0.003	0.009						
Cd								
Pb	0.03	0.034	0.08		0.22		0.11	
Total	63.6	64.5	38.0	26.9	67.7	0.51	68.0	1.69

These results showed that extracts of fumes from the welding of stainless steel, i.e., from operations number 1 to 4 as shown above, are mutagenic, whereas those from mild steel (operations 5 and 6) are not. Furthermore, particles produced by welding stainless steel using the FCAW method are more mutagenic than those obtained by using the SMAW method. On the other hand, particles in the fume produced by SMAW welding under short arc transfer conditions are more mutagenic than those produced by spray arc transfer.

In an attempt to explain why fumes produced by the FCAW method are more mutagenic than those produced by SMAW, Stern (Refs. 33 and 34) speculated that the particle size as well as the water solubility of these fumes might play a role. Analysis of particle size showed that particulates from SMAW fumes are generally smaller than those from FCAW fumes, having mass median diameters of 0.25 and 0.50 microns, respectively. This means that higher pulmonary retention of SMAW fumes would be expected. The water-soluble fraction from the FCAW process is 40 to 60 percent; that from the SMAW process averaged 0.2 to 1.5 percent. These results show that mutagenicity of the welding fume is a function of the material as well as the welding method. Although these results proved beyond doubt the mutagenic effects of extracts of fume from welding stainless steel, their impact on the welder's health cannot be assessed. However, as a precautionary measure, the concentrations of the fumes generated from welding stainless steel should be diminished to a minimum, especially when the FCAW technique is used.

As is the case of experiments on laboratory animals, mutagenicity studies using only one or more components of the welding fumes were performed. The analysis and interpretation of such results should be done with great care.

An *in vitro* study by Voroshilin et al. (Ref. 184) using human leukocytes culture showed that the hydrated acetates of zinc, cadmium, and cobalt are capable of inducing a mutagenic effect when used in small concentration. In another investigation, Voroshilin et al. (Ref. 184) studied the mutagenic effects of various inorganic compounds. They found that the inhalation of zinc oxide increased the frequency of chromosome abnormalities in bone marrow cells of albino rats.

References

1. Villaume, J.E.; Wasti, K.; Liss-Suter, D.; and Hsiao, S. 1978. *Effects of welding on health.* Miami, FL: American Welding Society.

2. Sunderman, F.W., and Kincaid, J.F. 1954. Nickel poisoning studies on patients suffering from acute exposure to vapors of nickel carbonyl. *J. Am. Med. Assoc.* 155:889-894.

3. Norseth, T., and Gundersen, N. 1978. Urinary excretion of nickel as an index of nickel exposure. *Ann. Clin. Lab. Sci.* 8:498.

4. Tola, S.; Kilpio, J.; Virtamo, M.; and Kaapa, K. 1977. Urinary chromium as an indicator of the exposure of welders to chromium. *Scand. J. Work Environ. and Health* 3:192-202.

5. Gylseth, G.; Gundersen, N.; and Langard, S. 1977. Evaluation of chromium exposure based on a simplified method for urinary chromium determination. *Scand. J. Work Environ. and Health* 3:28-31.

6. Stanescu, D.; Veriter, C.; Frans, A.; Goncette, L.; Roels, H.; Lauwerys, R.; Brasseur, L. 1977. Effects on lung of chronic occupational exposure to cadmium. *Scand. J. Resp. Dis.* 58:289-303.

7. Friberg, L. 1957. Proteinuria in chronic cadmium poisoning after comparatively short exposure to cadmium dust: a report of two cases. *A.M.A. Arch. Ind. Health* 16:30-1.

8. Smith, J.C., and Kench, J.E. 1957. Observations on urinary cadmium and protein excretion in men exposed to cadmium oxide dust and fume. *Brit. J. Ind. Med.* 14(4):240-245.

9. Cramer, K.; Goyer, R.A.; and Wilson, H. 1973. Renal ultrastructure, renal function and parameters of lead toxicity in workers with different lengths of lead exposure. *Environ. Health Aspects* 463-472.

10. Kehoe, R.A. 1963. Industrial hygiene and toxicology. In *Inter-Science Publ.,* ed. F.A. Patty, p. 955. New York.

11. Schrenk, H.H., and Schreibers, L., Jr. 1958. Urinary arsenic levels as an index of industrial exposure. *Amer. Ind. Hyg. Assoc. J.* 19(3):225-228.

12. Largent, F.J. 1961. *Fluorosis, the health aspects of fluorine compounds.* Columbus, OH: Ohio State Univ. Press.

13. Simon, J.H. 1965. In *Fluorine Chemistry,* vol. 4. New York: Academic Press Inc.

14. Jolly, S.S.; Prasad, S.; and Sharma, R., et al. 1970. Endemic fluorosis in India. *J. Assoc. Physicians India* 18:459-71.

15. Glover, J.R. 1967. Selenium in human urine: a tentative maximum allowable concentration for industrial and rural populations. *Ann. Occup. Med.* 10:3-14.

16. Ulfvarson, U., and Wold, S. 1977. Trace-element concentrations in blood samples from welders of stainless steel or aluminum and a reference group. *Scand. J. Work Environ. and Health* 3:183-191.

17. Jaremin, B.; Chmielewski, J.; Dolmierski, R.; Kwiatkowski, S.R.; and Szczekocki, W. 1975. A case of zinc fever in the course of a brain tumor. *Bull. Inst. Mar. Trop. Med. Gdynia* 26:323-328.

18. Clausen, J., and Rastogi, S.C. 1977. Heavy metal pollution among autoworkers: I. Lead. *British Journal of Industrial Medicine* 34:208-215.

19. Clausen, J., and Rastogi, S.C. 1977. Heavy metal pollution among autoworkers: II. Cadmium, chromium, copper, manganese, and nickel. *British Journal of Industrial Medicine* 34:216-220.

20. Fishbein, L. 1973. Nitrogen gas (oxides of nitrogen, ammonia) in *Chromatogr. Environ. Hazards* 2:237-257. New York: Elsevier Scientific Publishing Company.

21. Tsalev, D.L.; Langmyhr, F.J.; and Gunderson, N. 1977. Direct atomic absorption spectrometric determination of manganese in whole blood of

unexposed individuals and exposed workers in a Norwegian manganese alloy plant. *Bulletin of Environmental Contamination and Toxicology* 17(6): 660-666.

22. Parker, M.M.; Humoller, F.L.; and Mahler, D.J. 1967. Determination of copper and zinc in biological materials. *Clin. Chem.* 13:40-48.

23. Cotzias, G.C.; Miller, S.T.; and Edwards, J. 1966. Neutron activation analysis: the stability of manganese concentration in human blood and serum. *J. Lab. Clin. Med.* 67:836-849.

24. Kubota, J.; Lazar, V.A.; and Losee, F. 1968. Copper, zinc, cadmium, and lead in human blood from 19 locations in the United States. *Arch. Environ. Health* 16:788-793.

25. Allaway, W.H.; Kubota, J.; Losee, F.; and Roth, M. 1968. Selenium, molybdenum and vanadium in human blood. *Arch. Environ. Health* 16:342-348.

26. Thoburn, T.W., and Larsen, L.B. 1976. Health hazard evaluation determination report no. 74-16-272, Marathon Battery Company, Waco, Texas. Cincinnati, OH: National Institute for Occupational Safety and Health.

27. Sunderman, F.W. 1980. The toxicology and metabolism of nickel compounds. *Toxicology Research Project Directory 5.*

28. DeBruin, A. 1980. *Biochemical Toxicology of Environmental Agents.* Elsevier/North-Holland Biomedical Press.

29. Jindrichova, J. 1969. Anwendungsmoglichkeit der Mangandestimmung im Stuhl als Expositionstest. *Int. Arch. Gewerbe Path.* 25:347-59.

30. Kobayashi, H.; Makamura, Y.; and Ishiyama, K. 1978. Measurements of fume produced during CO_2 gas welding. *Kikuzo Kikai Gijustsu* 32:26-35.

31. Kobayashi, M.; Maki, S.; Hashimoto, Y.; and Saga, T. 1978. Some considerations about the formation mechanism of welding fumes. *Welding in the World* 16:238-249.

32. Stern, R.M. 1976. The production and characterization of a reference standard welding fume. Part one of report SF 7605 to The Danish Welding Institute.

33. Stern, R.M. 1977a. Publication 77.05, *A chemical, physical and biological assay of welding fume, part I: Fume characteristics.* Copenhagen: The Danish Welding Institute.

34. Stern, R.M. 1977b. Publication 77.08, *A chemical, physical and biological assay of welding fume, part II: Implications of positive mutagenic tests.* Copenhagen: The Danish Welding Institute.

35. Stern, R.M. 1978. Risks in chromium and nickel exposure for stainless steel welders (meeting abstract). *Ann. Clin. Lab. Sci.* 8:501.

36. Magnusson, J. 1978. The problems of fumes in welding. *ESAB Svetsaren* 1:8-17.

37. Hinrichs, J.F. 1978. Radiation and ore welding: new data to enhance safety. *Welding Design and Fabrication,* March:102-103.

38. Zaborski, L. 1976. Studies of the visible spectrum of the welding arc. *Bull. Inst. Mar. Trop. Med. Gdynia* 27:267-277.

39. Tierney, M.P. 1977. Analysis of mine injuries associated with maintenance and repair in metal and nonmetal mines. U.S. Dept. of The Interior, Mining Enforcement and Safety Administration.

40. Guidotti, T.L.; and DeNee, P.B. 1978. Arc welders' pneumoconiosis: application of advanced scanning electron microscopy. *Arch. Environ. Health* May/June:117-124.

41. Eskilsson, C.; Ekelund, S.; and Werlefors, T. 1977. Welding fume characterization by computer controlled SEM. Report IM-1244 to the Swedish Institute for Metal Research.

42. Patel, K.C.; Sheth, S.M.; and Kamat, S.R. 1977. Arc welder's lung (a case report). *Journal of Postgraduate Medicine* 23:35-38.

43. Stettler, L.E.; Groth, D.H.; and Mackay, G.R 1977. Identification of stainless steel welding fume particulates in human lung and environmental samples using electron probe microanalysis. *Am. Ind. Hyg. Assoc. J.* 38:76-82.

44. Ross, D.S. 1978. Welders' health—non-respiratory aspects. *Metal Construction* 5:204-409.

45. Singh, H., and Gupta, V.P. 1977. Investigation of health hazards of low hydrogen electrode. *Environ. Pollut. Hum. Health. Proc. Int. Symp.* pp. 202-210.

46. Gola, A., and Galazka, Z. 1978. Health status of a group of factory arc welders. *Medycyna Pracy* 29:163-168.

47. Rai, T. 1978. The haemolytic action of particulate metals (Cd, Cr, Co, Fe, Mo, Ni, Ta, Ti, Zn, Co-Cr alloy). *J. Path.* 125:81-89.

48. Raynor, A.C.; Vargas-Cortes, F.; Alexander, R.W.; and Bingham, H.G. 1979. Clear-cell sarcoma with melanin pigment: a possible soft-tissue variant of malignant melanoma. *Journal of Bone and Joint Surgery* 61-A:276-380.

49. Doll, R.; Mathews, J.D.; and Morgan, L.C. 1977. Cancers of the lung and nasal sinuses in nickel workers: a reassessment of the period of risk. *British Journal of Industrial Medicine* 34:102-105.

50. Milham, S., Jr. 1976. Cancer mortality patterns associated with exposure to metals. *Annals of the New York Academy of Sciences* 271:243-249.

51. Pantucek, M.B. 1975. Hygiene evaluation of exposure to fluoride fume from basic arc welding electrodes. *Ann. Occup. Hyg.* 18:207-212.

52. Buncker, C.R.; Emmett, E.A.; and Suskind, R.R. 1977. An epidemiologic survey of welders (abst.). *Am. J. Epidemiol.* 106:243.

53. Forrest, M.E.; Skerker, L.B.; and Nemiroff, N.J. 1978. Hard metal pneumoconiosis: another cause of diffuse interstitial fibrosis. *Radiology* 128:609-612.

54. Maxild, J.; Anderson, M.; and Kiel, P. 1978. Mutagenicity of fume particles from metal arc welding on stainless steel in the salmonella/microsome test. *Mutation Research* 56:235-243.

55. Buckle, E.R. 1978. Nature and origin of particles in condensation fume: a review. *Journal of Microscopy* 114(2):205-214.

56. Kayano, T.; Sato, S.; and Morishita, Y. 1978. Measurement of welding fume, ozone and nitric oxide concentration in aluminum arc welding. *J. Light Met. Weld.* 16:266-272.

57. Senn, T.J. 1977. Evaluation of the hazard associated with fabricating beryllium copper alloys. U.S. Dept. of Commerce *MTIS*, UCRL-52258.

58. Zimmerman, N.; Drehmel, D.C.; and Abbott, J.H. 1978. Characterization and generation of metal aerosols: a report to EPA-600/7-78-613. Environmental Protection Agency.

59. Walser, H., and Frei, A. 1977. Studies and experiences concerning ozone development deriving from inert gas welding of aluminum. *DVS Berichte* 45:40-46.

60. Opekar, B., and Kubovec, G. 1975. Pracovni prostredi pri svarovani a rezani hliniku. *Pracov. Lek.* 27:165-170.

61. Baum, K. 1978. Ermitteln des Ozongehalts beim Wolfram-InertgasschweiBen mit Analysengerat und Prufrohrchen – *Vergleich der McBverfrahren. SchweiBen und Schneiden* 30:439-441.

62. Tabershaw, I.R.; Utidjian, H.M.D.; and Kawahara, B.L. 1977. Chemical hazards. In *Occupational diseases, a guide to their recognition.* NIOSH publication No. 77-181, pp. 131-442.

63. Press, H. 1977. *Conclusions applicable to actual practices from measurements of noxal concentrations during welding with shielded rod electrodes in small areas.* IIW-Doc. VIII-782-78. Berlin: Federal Institution of Materials Testing.

64. Moss, E.; Murray, W.; Parr, W.; and Conover, D. 1977. Physical hazards. In *Occupational diseases, a guide to their recognition.* Department of Health, Education and Welfare. Publication No. 77-181, pp. 467-496. Cincinnati, OH: National Institute for Occupational Safety and Health.

65. Moss, C.E. 1979. Optical radiation transmission levels through transparent welding curtains. *Welding Research Supplement* March:69-S-75-S.

66. Ingram, J.W., and Horstman, S.W. 1977. A field study of near ultraviolet welding irradiance. *Am. Ind. Hyg. Assoc. J.* 38:461-466.

67. Kuzina, A.S. 1978. Hygienic characteristics of heat emission of the welding arc in some kinds of coiler metal welding. *Vrach. Delo.* 9:108-110.

68. Johnson, P., and Shipps, G. 1977. Survey report, Corpus Christi Army Depot, Corpus Christi, Texas. Cincinnati, OH: National Institute for Occupational Safety and Health.

69. Johnson, P., and Roberts, D. 1978. Survey report PB-291-678 of General Motors Assembly Division, Norwood, Ohio. Cincinnati, OH: National Institute for Occupational Safety and Health.

70. Hoecker, J.E.; Durkin, P.R.; Hanchett, A.; Davis, L.N.; Meyland, W.M.; and Bosch, S.J. 1977. Information Profiles on Potential Occupational Hazards. *NIOSH.* October:1-330.

71. Ross, D.S. 1977a. Welders' health – a historical review. *Metal Construction* 7:303-304.

72. Ross, D.S. 1977b. Welders' health – a review of currently available information. *Metal Construction.*

73. Sloan, D. 1978. The health hazards of welding, part 1. *Occupational Safety and Health* 8:28-29.

74. Naumenko, I.M. 1977. Low temperatures as an occupationally adverse factor in work of electric welders. *Gig. Tr. Prof. Zabol* 3:15-18.

75. United Kingdom Health and Safety. 1978. *Executive Guidance Note MSIS.* London: H.M. Stationary Office.

76. Clapp, D.E., and Owen, R.J. 1977. An investigation of potential health hazards of arc welding fume growth with time. *Welding Research Supplement* December:380s-385s.

77. Brown, K.L. 1979. Comments concerning the proposed rulemaking on the use of asbestos fibers. Docket #OTS 61005: a letter to Office of Toxic Substances (TS-793), 7 December, 1979.

78. Task Group on Lung Dynamics. 1966. Deposition and retention models for internal dosimetry of the human respiratory tract. *Health Physics* 12:173.

79. Wolf, A.F. 1975. Occupational diseases of the lungs. *Annals of Allergy* 35:165-171.

80. Werner, U. 1977. Diseases of the upper respiratory tract in welders. *Z. Gesamte Hyg.* 23:731-734.

81. Hanson, R.R., and Kasik, J.E. 1977. The pneumoconioses. *Heart and Lung* 6:646-652.

82. Fabre, M.; Marais, M.; Ladouch, A.; Paillas, J.; Blanchon, P.; and Quillard, J. 1976. La pneumonoconiose des soudeurs a l'arc. *Arch. Anat. Cytol. Path.* 24:281-485.

83. Irmscher, G.; Beck, B.; Ahlendorf, W.; Anspach, M.; Konetzke, G.; Ludwig, V.; and Sturm, W. 1975. Experience in assessment of doubtful cirrhosis of the lung caused by welding smoke. *Z. Gesamte Hyg. Grenzgeb.* 21:562-566.

84. Roussell, D., and Pottier, M. 1978. Un cas de berylliose pulmonaire chez un a soudeur a l'arc. *Arch. Mal. Prof. Med. Trav. Secur.* pp. 85-87.

85. Sprince, N.L.; Kanarek, D.J.; Weber, A.L.; Chamberlin, R.I.; Kazemi, H. 1978. Reversible respiratory disease in beryllium workers. *Amer. Review of Resp. Dis.* 117:1011-1017.

86. Prowse, K. 1977. Nitrous fume poisoning. *Bull. Eur. Physiopath. Resp.* 13:191-203.

87. Kurta, A.V. 1976. Acute pulmonary oedema due to welding in a confined area (abst.). *Aust.-N. Z. J. Med.* 6:369.

88. Hocking, B. 1976. Welder's lung. *Papua New Guinea Medical Journal* 19:108-110.

89. Lehnert, G.; Morgan, A.; Szadkowski, D.; and Zielhuis, R.L., eds. 1974. Halogenated hydrocarbon solvents: long-term effects and biological sampling in human beings. *Int. Arch. Arbeistmed.* 33:251-255.

90. Mosekilde, L., and Pallisgaard, G. 1977. Alveolitis fibrosa – an occupational disease? *Scand. J. Resp. Dis. Supp.* 99:43.

91. Marek, K., and Kujawska, A. 1975. Evolution of functional respiratory disorders in different types of pneumoconiosis. *Bull. Physiopath. Resp.* 11:597-610.

92. Spacilova, M.; Navratil, M.; and Koval, Z. 1977. A contribution to the assessment of pulmonary changes in arc welders. *Pracov. Lek.* 29:51-53.

93. Bergert, K.D.; Krause, M.; and Mahrlein, W. 1978. Hinweise uber den Einflud einer langjahrigen SchweiBertatigkeit auf die Entwicklung eines chronisch obstruktiven Atemwegssyndroms. *Z. Ges. Inn. Med., Jahrg.* 33:799-800.

94. Oxhoj, H.; Bake, B.; Wedel, H.; Wilhelmsen, L. 1978. Effects of electric arc welding on ventilat lung function. *Bull European de Physiopatholo Respiratoire* 14:48-49.

95. Kalliomaki, P.L.; Alanko, K.; Korhonen, O.; Ma son, T.; Vaaranen, V.; and Koponen, M. 19 Amount and distribution of welding fume l contaminants among arc welders. *Scand. J. W Environ. and Health* 4:122-130.

96. Kalliomaki, P.L.; Korhonen, O.; Vaaranen, Kalliomaki, K.; and Koponen, M. 1978. Lung tention and clearance of shipyard arc weld *Int. Arch. Occup. Environ. Health* 42:83-90.

97. Kalliomaki, P.L.; Karp, P.J.; Katila, T.; Makip P.; Saar, P.; and Tossavainen, A. 1976. Magne measurements of pulmonary contamination. *Sca J. Work Environ. and Health* 4:232-239.

98. Records, R.E. 1979. Primary care of ocu emergencies, part 2: thermal, chemical, and n traumatic eye injuries. *Postgraduate Medicine* 157-163.

99. Zaborski, L. 1978. Dazzle at welding. *Bull. In Mar. Trop. Med.* 29:199-209.

100. Krasniuk, E.P.; Fleishman, F.M.; Medvedovska C.P.; Zaritskaya, L.A.; Pashchenko, V.A.; Karulina, I.S. 1976. Effect of argon arc welding the health status of workers. *Vrach. Delo.* 3:1 142.

101. Tengroth, B., and Vulcan, J. 1977. Welding lig *Strahlentherapie* 153:267-272.

102. NIOSH. 1972. *Criteria for a recommended st dard, occupational exposure to ultraviolet rad tion.* Cincinnati, OH: Department of Heal Education and Welfare.

103. NIOSH. 1978. *Safety and health in arc weld and gas welding and cutting,* publication 78-1 Cincinnati, OH: Department of Health, Edu tion and Welfare.

104. Kasansky, O. 1978. Personal protection from n ionizing radiation in welding – a technical vi *Australian Welding Journal* July/August:16-18.

105. Tengroth, B. 1976. Safety glasses in weldi *Environmental Research* 11:283-284.

106. Crow, M.L., and McCoy, F.J. 1977. Volume crease by Z-plasty to the finger skin: its appl tion in electrical ring burns. *The Journal of Ha Surgery* 2:402-403.

107. Jirasek, L. 1978. Occupational exogenous sider of the skin. *Cs. Derm.* 53:248-251.

108. Rogowsky, M.; Putten, M.V.; and Baudoux, 1978. Etude critique de la formation de carbo hemoglobine (HbCo) chez les soudeurs travail

en atmosphere a faible concentration d'oxyde de carbone (CO). *Arch. Mal. Prof.* 39:151-161.

109. Bauer, S.; Sofronievska, Z.; and Orusev, T. 1977. Occupational exposure to carbon monoxide of the welding workmen in the industry of metal preparation. *God. Zb Med. Fak. Skopje* 23:241-244.

110. Platonov, A.A. 1978. Features distinguishing the mental activity in patients with chronic manganese poisoning. *Gig. Tr. Prof. Zabol.* 10:23-26.

111. Kominsky, J.R., and Tanaka, S. 1976. Health hazard evaluation determination report number 76-16-336, Abex Corporation, Pueblo, Colorado. Cincinnati, OH: National Institute for Occupational Safety and Health.

112. Avakyan, M.A.; Nargizyan, G.A.; Aghabalyan, E.A.; and Alexanyan, A.S. 1978. Some biochemical indexes in workers in contact with industrial dust containing molybdenum. *Zh. Eksp. Klin. Med.* 18:69-74.

113. National Safety Council. 1979. Data sheet cadmium (Cd). *National Safety News* January:63-64.

114. Kjellstrom, T.; Evrin, P.E.; and Rahnster, B. 1977. Dose-response analysis of cadmium-induced tubular proteinuria. *Environmental Research* 13: 303-317.

115. British Occupational Hygiene Society Committee on Hygiene Standards; Subcommittee on Cadmium. 1977. Hygiene standard for cadmium. *Ann. Occup. Hyg.* 20:215-228.

116. Hunter, D. 1974. *The Diseases of Occupations.* Oxford: The English Universities Press, Ltd.

117. Wintrobe, C.E., ed. 1974. *Harrison's Principals of Internal Medicine.* 7th ed. New York: McGraw-Hill.

118. Morgan, W.K., and Seaton, A. 1975. *Occupational Lung Diseases* pp. 321-345. Philadelphia: W.B. Saunders.

119. Gun, R.T. 1977. Febrile episodes in welders: relationship to occupational exposure and smoking habits. *Community Health Studies* 1:22-25.

120. Hopper, W.F. 1978. Case report – metal fume fever. *Postgraduate Medicine* 63:123-127.

121. Dula, D.J. 1978. Metal fume fever. *JACEP* 7: 448-450.

122. Waldbott, G.L. 1978. Fever-producing agents. In *Health effects of environmental pollutants* 2d ed., pp. 123-129.

123. Ferris, W.L. 1977. The welding of zinc coated steels. *Corrosion and Coatings South Africa,* August:17.

124. Calnan, C.D. 1979. Metal fume fever. *Contact Dermatitis* 5:125-126.

125. Cantrell, A.D.; Kilroe-Smith, A.; Simoes, N.N.; and Border, E.A. 1977. The effect of zinc and pH on the behaviour of gamma-aminolevulinic acid dehydratase activity in baboons exposed to Uad. *Br. J. Indus. Med.* 34:110-113.

126. Beretta, F.; Cortona, G.; and De Bigontina, G. 1977. Hazard of lead absorption in female workers employed on microweldings. *Med. Lavoro* 68: 340-348.

127. Gramlich, J.W.; Nachlan, L.A.; Murphy, T.J.; and Moore, L.J. 1977. The determination of zinc, cadmium and lead in biological and environmental materials by isotope dilution mass spectrometry. *Trac. Subs. Environ. Health* 11:376-380.

128. Emmett, E.A. 1975. Occupational skin cancer: a review. *J. of Occup. Med.* 17:44-49.

129. Barton, R.T. 1977. Nickel carcinogenesis of the respiratory tract. *Journal of Otolaryngology* 6: 412-422.

130. New York Academy of Sciences. 1977. *Cancer and the worker.* ISBN 0-89072-058-4.

131. Howe, G.R.; Chambers, L.; Gordon, P.; Morrison, B.; and Miller, A.B. 1977. An epidemiological study of bladder cancer (abst.). *Am. J. Epidemiol.* 106:239.

132. Levy, B.S.B., and Meyer, C.R. 1977. Health hazard evaluation determination report no. 76-1-388, Bohn Aluminum and Brass Corporation, Danville, Illinois. Cincinnati, OH: National Institute for Occupational Safety and Health.

133. Goldsmith, J.R., and Guidotti, T.L. 1977. Environmental factors in the epidemiology of lymphosarcoma. *Pathol. Ann.* 12:411-425.

134. Pikulhskaya, A.F., and Gulhko, S.N. 1975. Insufficiency of the external respiration function in electric welders. *Vrach. Delo.* 1:127-130.

135. Hewitt, P.J.; Hicks, R.; and Lam, H.F. 1978. The generation and characterization of welding fumes for toxicological investigations. *Ann. Occup. Hyg.* 21:159-167.

136. Hewitt, P.J., and Hicks, R. 1972. Neutron activation analysis of blood and body tissue from rats exposed to welding fume. In proceedings of a symposium *Nuclear activation techniques in the life sciences.* 1972, ed. N. Kruppner, pp. 219-232. Vienna: International Atomic Energy Agency.

137. Senczuk, W., and Mlynarczyk, W. 1978. Investigations of the correlation between changes in lungs

caused by industrial dusts and the composition of amino acids in blood plasma. *Medycyna Pracy* 29:365-370.

138. Doi, K. 1978. An experimental study on the biological effects of electric arc welding fumes containing antimony. *Jap. J. Ind. Health* 20:9-23.

139. Hicks, R.; Hewitt, P.J.; and Lam, H.F. 1979. An investigation of the experimental induction of hypersensitivity in the guinea pig by material containing chromium, nickel and cobalt from arc welding fumes. *Int. Archs. Energy Appl. Immun.* 59:262-265.

140. Skilleter, D.N., and Paine, A.J. 1979. Relative toxicities of particulate and soluble forms of beryllium to a rat liver parenchymal cell line in culture and possible mechanisms of uptake. *Chem.-Biol. Interactions* 24:19-33.

141. Skilleter, D.N., and Price, R.J. 1978. The uptake and subsequent loss of beryllium by rat liver parenchymal and non-parenchymal cells after the intravenous administration of particulate and soluble forms. *Chem.-Biol. Interactions* 20:383-396.

142. Conklin, A.W.; Hadley, J.G.; and Sanders, C.L. 1978. Systemic toxicity of inhaled cadmium oxide in rats (abst.) *Pharmacologist* 20, 212.

143. Gardner, D.E.; Miller, F.J.; Illing, J.W.; and Kirtz, J.M. 1977. Alterations in bacterial defense mechanisms of the lung induced by inhalation of cadmium. *Bull. Europ. Physiopath. resp.* 13:157-174.

144. Asvadi, S., and Hayes, J.A. 1978. Acute lung injury induced by cadmium aerosol II. Free airway cell response during injury and repair. *American Journal of Pathology* 90:89-98.

145. Bouley, G.; DuBreuil, A.; Despaux, N.; and Boudene, C. 1977. Toxic effects of cadmium microparticles on the respiratory system. *Scand. J. Work Environ. and Health* 3:116-121.

146. Hayes, J.A.; Asvadi, S.; Strauss, R.H.; and Palmer, K.C. 1977. Alveolar epithelial repair and changes in free airway cell populations following cadmium injury. *ERDA Symp. Ser.* 43:582-589.

147. Koshi, K.; Homma, K.; and Sakabe, H. 1978. Damaging effect of cadmium oxide dust to the lung and its relation to solubility of the dust. *Industrial Health* 16:81-89.

148. Boisset, M.; Girard, F.; Godin, J.; and Boudene, C. 1978. Toxicologie: Kinetics of the pulmonary clearance of inhaled cadmium and of its accumulation in liver and kidneys of the rat. *C.R. Acad. Sc. Paris* 287:61-64.

149. Sanders, C.L.; Hadley, J.G.; Conklin, A.W.; an Adee, R.R. 1978. Antagonism of cadmium-induce pulmonary toxicity by WR2721. *Toxicology Let ters* 2:323-328.

150. Prigge, E.; Baumert, H.P.; and Muhle, H. 197 Effects of dietary and inhalative cadmium o hemoglobin and hematocrit in rats. *Bull. Enviro Contam. and Toxic.* 17:585-590.

151. Prigge, E. 1978. Early signs of oral and inhalativ cadmium uptake in rats. *Arch. Toxicol.* 40:23 247.

152. Prigge, E. Inhalative cadmium effects in pregnan and fetal rats. *Toxicology* 10:297-309.

153. Bus, J.S.; Vinegar, A.; and Brooks, S.M. 197 Biochemical and physiologic changes in lungs o rats exposed to a cadmium chloride aerosol. *Ame Rev. Resp. Dis.* 118:573-580.

154. Hamilton, A., and Hardy, H.L. 1974. *Industri Toxicology,* 3rd. ed. Acton, MA: Publishin Sciences Group.

155. Wehner, A.P.; Busch, R.H.; Olson, R.J.; an Craig, D.K. 1972. Chronic inhalation of coba oxide and cigarette smoke by hamsters. *Am. In Hyg. Assoc. J.* 38:338-346.

156. Georgiadi, G.A., and Elkind, L.A. 1978. Morph logical changes in respiratory tract mucosa und the effect of a metallic cobalt aerosol under chron experimental conditions. *Zh. Ushn., Gorl. Bolez* 3:41-45.

157. Georgiadi, G.A. 1978. Change of some deh drogenases and non-specific enzymes activity respiratory tract mucosa of rats under the influen of metal cobalt powder in chronic experimen *Zh. Ushn., Nos. Gorl. Bolezn.* 1:63-67.

158. Zaidi, S.H.; Dogra, R.S.; Sahu, A.P.; Khanna, S and Shanker, R. 1977. Experimental infecti pneumoconiosis: effect of candida albicans a iron ore dust on the lungs of guinea pigs. *Industr Health* 15:123-129.

159. Levens, D.; Nicholas, H.; Brain, J.; and Huber, 1977. Impairment of intrapulmonary anti-bacter defenses following exposure to aerosolized ine particles (abst.). *Clinical Research* 25:419a.

160. Kavet, R.I.; Brain, J.D.; and Levens, D.J. 197 Characteristics of pulmonary macrophages lavag from hamsters exposed to iron aerosols. *Laborato Investigation* 38:312-319.

161. Watson, A.Y., and Brain, J.D. 1979. Uptake iron oxide aerosols by mouse airway epitheliu *Laboratory Investigation* 40:450-459.

162. Quan, S.G., and Golde, D.W. 1977. Surfa

morphology of the human alveolar macrophage. *Experimental Cell Research* 109:71-77.

63. Oberdorster, G.; Gibb, F.R.; Beiter, H.; Lu, S.T.; and Morrow, P.E. 1978. Studies of the lymphatic drainage of dog lungs. *J. Toxicol. Environ. Health* 4:571-586.

64. Singh, J.; Kaw, J.L.; and Zaidi, S.H. 1977. Early biochemical response of pulmonary tissue to manganese dioxide. *Toxicology* 8:177-184.

65. Griffin, T.B., and Coulston, F. 1978. Inhalation toxicology of airborne particulate manganese in rats and rhesus monkeys (abst.). *Toxicol. Appl. Pharmacol.* 45:290.

66. Ulrich, C.E.; Rinehart, W.; Busey, W.; and Dorato, M.A. 1979. Evaluation of the chronic inhalation toxicity of a manganese oxide aerosol, part 2: Clinical observations, hematology, clinical chemistry and histopathology. *Am. Ind. Hyg. Assoc. J.* 40:322-329.

67. Bergstrom, R., and Rylander, R. 1977. Pulmonary injury and clearance of MnO_2 particles. *ERDA Symp. SER.* 3:523-32.

68. Camner, P.; Johansson, A.; and Lundborg, M. 1978. Alveolar macrophages in rabbits exposed to nickel dust - ultrastructural changes and effect on phagocytosis. *Environmental Research* 16:226-235.

9. Adalis, D.; Gardner, D.E.; Miller, F.J.; and Coffin, D.L. 1977. Toxic effects of cadmium on ciliary activity using a tracheal ring model system. *Environmental Research* 13:111-120.

0. Hopfer, S.M.; Saunderman, F.W.; Fredrickson, T.N.; and Morse, E.E. 1978. Nickel-induced erythrocytosis: efficacies of nickel compounds and susceptibilities of rat strains. *Annals of Clinical and Laborary Science* 8:386-402.

1. Hopfer, S.M., and Sunderman, F.W. 1978. Manganese inhibition of nickel subsulfide induction of erythrocytosis in rats. *Research Communications in Chemical Pathology and Pharmacology* 19: 337-345.

2. Mathur, A.K.; Chandra, S.V.; Behari, J.; and Tandon, S.K. 1977. Biochemical and morphological changes in some organs of rats in nickel intoxication. *Arch. Toxicol.* 37:159-164.

3. Graham, J.A.; Miller, F.J.; Daniels, M.J.; Payne, E.A.; and Gardner, D.E. 1978. Influence of cadmium, nickel, and chromium on primary immunity in mice. *Environmental Research* 16:77-87.

174. Charles, J.N.; Gardner, D.E.; Coffin, D.L.; and Menzel, D.B. 1977. Augmentation of sulfate ion absorption from the rat lung by heavy metals. *Tox. Appl. Pharm.* 42:531-538.

175. Ishinishi, N.; Kodama, Y.; Nobutomo, K.; and Hisanaga, A. 1977. Preliminary experimental study on carcinogenicity of arsenic trioxide in rat lung. *Environ. Health Perspectives* 19:191-196.

176. Saknyn, A.V., and Blokhin, V.A. 1978. The induction of malignant tumors in rats by nickel-containing aerosols. *Vopr. Onkol.* 24:44-48.

177. NIOSH. 1979. *Registry of Toxic Effects of Chemical Substances 1978.* Cincinnati, OH: Department of Health, Education and Welfare.

178. Higginson, J. 1969. Present trends in cancer epidemiology. *Canadian Canc. Conf.* 8:40-75.

179. Epstein, S.S. 1974. Environmental determinants of human cancer. *Cancer Res.* 34:2425-2435.

180. Cairns, J. 1975. The cancer problem. *Scientific American* 233:63-78.

181. Purchase, I.F.H.; Longstaff, E.; Ashby, J.; Styles, J.A.; Anderson, D.; Lefevre, P.A.; and Westwood, F.R. 1976. Evaluation of six short-term tests for detecting organic chemical carcinogens and recommendations for their use. *Nature* 264:624-627.

182. De Serres, F.J. 1976. Mutagenicity of chemical carcinogens. *Mutation Res.* 41:43-50.

183. Bigaliev, A.B.; Elemesova, M.S.H.; and Turebaev, M.N. 1977. Evaluation of the mutagenic activity of chromium compounds. *Gig. Tr. Prof. Zabol.* 6:37-40.

184. Voroshilin, S.I.; Plotko, E.G.; Fink, T.V.; and Nikiforova, V.J. 1977. Cytogenetic effect of inorganic compounds of tungsten, zinc, cadmium, and cobalt on animal and human somatic cells. *Tsitol. Genet.* 2:241-243.

185. Bernacki, E.J., and Parsons, G.E. 1978. Urine nickel concentrations in nickel-exposed workers. *Am. Clin. Lab. Sci.* 8:498.

186. Wolf, A.F. 1977. Occupational lung disease, part 3: Conditions caused by noxious gases. *Consultant* 17:205-206.

Bibliography

lalis, D.; Gardner, D.E.; and Miller, F.J. 1978. ytotoxic effects of nickel on ciliated epithelium. *nerican Review of Respiratory Disease* 118:347-354.

ekseeva, I.S., and Norkin, Y.L. 1979. Hygienic evaluion of the environmental air in argon arc built-up elding of copper-nickel alloy with a consumable elecode. *Gig. Tr. Prof. Zabol.* 3:5-8.

herley, G.R.C. 1978. Health and safety at work, in ndrews D.R. (ed.): *Soldering - brazing, welding, and hesives.* London: Institution of Production Engineers.

bushkina, L.G., and Elnichnykh, L.N. 1977. Features culiar to changes in the lipids metabolism and oxyoline in the lungs of rats during different periods rking development of silicosis. *Gig. Tr. Prof. Zabol.* 43-46.

bushkina, L.G. 1977. The effect of different edible ions on the lipids metabolism in rats with experintal silicosis. *Vopr. Pitan.* 3:33-40.

letsky, V.S. 1977. Disturbance of free radicals processes the lungs during development of experimental silicosis. *g. Tr. Prof. Zabol.* 5:42-45.

oom, T.F., and Kramkowski, R.S. 1973. Health hazard aluation/toxicity determination report 73-154-146, her Body Division, G.M.C., Hamilton, Ohio. Cincinti, OH: National Institute for Occupational Safety d Health.

issett, M.; Girard, F.; Godin, J.; and Boudene, C. 78. Cadmium content of lung, liver, and kidney in s exposed to cadmium oxide fumes. *Int. Arch. Occup. viron. Hlth.* 41:41-53.

itish Standard Institution. 1977. Draft for developnt: Methods for the sampling and analysis of fume m welding and allied processes, part 1. Particulate tter. DD54:part 1.

own, J., and Robinson, C. 1978. Walk-through survey ort, American Cyanamid Co., Fibers Division, Santa sa Plant, Milton, Florida. Cincinnati, OH: National stitute for Occupational Safety and Health.

Cape, J.C. 1979. Health problems in motor vehicle manufacture. *Practitioner* 223, 73-78.

DeLong, W.R. 1978. OSHA regulatory proposals and philosophies pertaining to welding fumes. *Welding Journal* July:15-23.

Demmer, P., and Moll, D. 1978. Schadstoffarme Metallverar-beitung. *Umwelt* 8:54-56.

Dixon, R.D. 1978. Measurements of noise on the beam of an electron beam welding unit. *Welding Research Supplement* February: 51-S-56-S.

Elfimova, E.V.; Gusev, M.I.; Iudina, T.V.; Kotchetkova, T.A.; Smirnov, V.T.; and Titova, I.N. 1977. An experimental study of the action of low concentrations of nickel oxide on the body. *Gig. Sanit.* 12:18-22.

Frazier, M.E.; Andrews, T.K.; Thompson, B.B.; and Wincek, M.A. 1978. Toxicity of heavy metals in vitro (abst.). *Pharmacologist* 20:212.

Gilles, D. 1978. Health hazard evaluation determination report 77-124-479, Interlake Steel, Inc., Newport, Kentucky. Cincinnati, OH: National Institute for Occupational Safety and Health.

Gormley, I.P.; Wright, M.O.; and Ottery, J. 1978. The effect of toxic particles on the electro-physiology of macrophage membranes. *Ann. Occup. Hyg.* 21:141-149.

Grothe, I. 1977. Current problems of labor protection in the welding industry. *DVS Berichte* 45:21-25.

Gunter, B.J.; Ligo, R.; and Ruhe, R.L. 1976. Health hazard evaluation determination report number 75-30-266, Steiger Tractor, Inc. Fargo, North Dakota. Cincinnati, OH: National Institute for Occupational Safety and Health.

Gunter, B.J., and Shama, S. 1973. Health hazard evaluation/toxicity determination, Empire Detroit Steel Company, New Boston, Ohio. Cincinnati, OH: National Institute for Occupational Safety and Health.

Halton, N. 1977. Diesel and metal particulates on interferon synthesis. *Tox. Res. Proj. Dir. 4, No. 1.*

Morgantown, WV: National Institute for Occupational Safety and Health.

Hasegawa, N.; Hirari, A.; and Shibata, T., et al. 1970. Determination of lead in hair by atomic absorption spectroscopy for simple screen of lead intoxication. *Annu. Rep. Res. Inst. Environm. Med.* 18:1-5. Nagoya University.

Hervin, R.L.; Cormer, J.W.; and Butler, G.J. 1974. Health hazard evaluation/toxicity determination report 74-2/8-164, the Vendo Company, Kansas City, MO. Cincinnati, OH: National Institute for Occupational Safety and Health.

Holland, W.D. 1976. *Industrial hygiene survey of the Parahoe oil shale mining and processing facilities.* Denver, CO: US Dept. of Interior, Bureau of Mines.

Jaremin, B. 1977. Diagnostic usefulness of serum zinc concentration determinations in subjects with occupational exposure to zinc oxide vapours. *Poslki Tygodnik Lekarski* 32(26):997-1000.

Jenkins, N. 1975. Review of welding fume sampling and analysis. *Exploit. Weld. Prod. Technol. Int. Conf.* 1:88-95.

Jindrichova, J. 1978. Chromium-induced injuries in electric welders. *Z. Gesamte Hyg.* 24:86-88.

Jones, M. 1977. Industrial hygiene survey of Westinghouse Electric Corporation, Bloomington, Indiana. Cincinnati, OH: National Institute for Occupational Safety and Health.

Kinnersky, P. 1978. Health hazard guide. *New Civil Engineer* 320:24-31.

Kopito, L.; Byers, R.K.; and Schwachman, H. 1967. Lead in hair of children with chronic lead poisoning. *New Eng. J. Med.* 276:949-953.

Kramkowski, R.S., and Shama, S.K. 1973. Health hazard evaluation/toxicity determination, Modern Metal Products, Loves Park, Illinois. Cincinnati, OH: National Institute for Occupational Safety and Health.

Lam, H.F.; Hewitt, P.J.; and Hicks, R. 1978. The controlled generation of two types of welding fumes and their deposition and clearance in the rat and guinea pig (abst.). *Bradford University American Review of Respiratory Diseases* 117:246, 1978. Bradford, England.

Lange, A.; Smolik, R.; Zatonski, W.; Zwolinski, J.; Garncarek, D.; Okrojek, M.; and Gielgier, Z. 1977. Serum immunoglobulin and other protein levels in arc welders. *Int. Arch. Occup. Environ. Health* 38:189-196.

Larsen, L.B. 1973. Health hazard evaluation/toxicity determination, Hanford Foundry, San Bernadino, California. Cincinnati, OH: National Institute for Occupational Safety and Health.

Marek, K., and Kujawska, A. 1973. Effects of early pneumoconioses on respiratory function. *Bull. Phy path. Resp.* 9:1173-1187.

Markel, H.L., Jr. 1974. Health hazard evaluation/toxi determination report 73-111-125, Southern Steel Wire Company, Fort Smith, Arkansas. Cincinnati, National Institute for Occupational Safety and Healtl

Mastromatteo, E. 1977. *Fumes - effects, TLV's, and noise, heat and U.V. radiation: clinical effects tolerance levels.* Lecture presented before the Cana Welding Society, Toronto.

Molyneux, M.K.B. 1978. Hazards of welding recogni and hygiene standards. *CME* November:54-57.

Moseleu, C.L. 1977. Health hazard evaluation determ tion report no. 77-117-444, EAZ-Lift Towing Syst Elkhart, Indiana. Cincinnati, OH: National Institute Occupational Safety and Health.

Mudde, I.A. 1977. Vermelding van lasrookgegevens verpakking, voorstel tot een systeem van rookaanduid *Lastechniek* 43:199-202.

NIOSH. 1976. *Health and safety at work: in solder brazing, welding and adhesives.* Health hazard evaluat toxicity determination report 75-79-331, Health gineering Company, Fort Collins, Cincinnati, OH: tional Institute for Occupational Safety and Health.

NIOSH. 1977. *Health and safety guide for metal co and allied services.* Cincinnati, OH: National Institute Occupational Safety and Health.

NIOSH. 1978. Survey report of General Motors Assen Division, Norwood, Ohio. Cincinnati, OH: Nati Institute for Occupational Safety and Health.

Nishiyama, K.; Suzuki, Y.; Fujii, N.; Yano, H.; Ohr K.; and Miyai, T. 1977. Biochemical changes and ganese distribution in monkeys exposed to manga dioxide dust. *Tokushima J. Exp. Med.* 24:137-145.

Oakley, P.J. 1976. The problem of fumes in the we environment. *Welding Journal* June:465-472.

Okawa, M.T. 1977. Health hazard evaluation/tox determination report 77-24-366, Kaiser Marine Fac Vallejo, California. Cincinnati, OH: National Institut Occupational Safety and Health.

Oleinichenko, K.A. 1974. Health characteristics of ferent methods of welding aluminum bus-bars. *Proiz* 5:48-49.

Oliver, T.P.; and Molyneux, M.K.B. 1975. Calculatic risks in burning and welding. *Ann. Occup. Hyg.* 295-302.

Ostergaard, K. 1978. Renal cadmium concentratic relation to smoking habits and blood pressure. *Acta Scand.* 203:379-383.

Pantucek, M.B. 1976. Determination of fluoride urine as an exposure test welders with basic electr covni Lekarstvi 28:57-61.

rce, L.A. 1978. Systems design for safe control of lding fumes. *Pollution Engineering* July:51-53.

ov, L.N.; Kochetkova, T.A.; Gusev, M.I.; Markina, A.; Elfimova, E.V.; and Timonov, M.A. 1977. Accumu-on, distribution and morphological changes in the ly due to inhalation of metal cobalt aerosol. *Gig. it.* 6:12-15.

d, W.E., ed. 1978. *Inert gas welding (citations from NTIS data base).* Springfield, VA: National Technical ormation Service.

s, D.S. 1974. Welding fume and welders' health. *tal Construction* 6:55.

o, P. 1978. Urinary β_2 microglobulin related to onic low level exposure to metallic cadmium dust in nish cadmium processing workers. *Scand. J. Work iron. and Health* 4:131-136.

r, F.H. 1978. Air pollution exposures to five target th hazards. *Am. Ind. Hyg. Assoc.* 39:928-931.

kadze, V.P., and Vasilov, B.G. 1979. Occupational nchial asthma due to the effect of manganese. *Tr. Prof. Zabol.* 9:19-23.

imi, B.; Ziskind, M.; and Weill, H. 1978. The relation ilica dust to accelerated silicosis. *Ecotoxicology and ironmental Safety* 1:429-436.

ders, C.L. 1979. Alveolar clearance of metal oxides. *icol. Res. Proj. Dir.* 4:1-11.

lling, D. 1977. Arbeitshygienische Probleme beim SchweiBen. *ZIS-Mitteilungen* 6:722-730.

sintaffar, E.L., and Parr, W.H. 1975. Ultraviolet ation spectra and exposure levels from metal-inert (MIG) welders (abst.). In *Nineteenth Annual Meeting he Health Physics Society,* p. 634.

ikov, L.P. 1977. Safety orders and industrial sanita- during mechanical treatment and use of aluminum titanium alloys. Dept. of the Navy.

n, D. 1978. Skin diseases - how industrial workers vulnerable. *Welding and Metal Fabrication* 46(4): 276.

ght, F.Y. 1978. Letter from American Welding ety to OSHA. *Welding Journal* 7:22-23.

ub, W.E. 1977a. Health hazard evaluation determina- report no. 76-67, Fischer and Porter Company, minster, Pennsylvania. Cincinnati, OH: National In-te for Occupational Safety and Health.

ub, W.E. 1977b. Health hazard evaluation determina- report no. 77-30-405, Corpus Christi Army Depot, us Christi, Texas. Cincinnati, OH: National Institute Occupational Safety and Health.

Sullivan, J.F.; Blotcky, A.J.; Jetton, M.M.; Hahn, H.K.J.; and Burch, R.E. 1979. Serum levels of selenium, calcium, copper magnesium, manganese and zinc in various human diseases. *J. Nutr.* 109:1432-1437.

Ulrich, C.E.; Rinehart, W.; and Busey, W. 1979. Evaluation of the chronic inhalation toxicity of a manganese oxide aerosol, part 1: Introduction, experimental design, and aerosol generation methods. *Am. Ind. Hyg. Assoc. J.* 40:238-244.

Ulrich, L.; Malik, E.; Borovsky, M.; and Sulcova, M. 1977. Ucinok Novych zvaracich postupov na zdravie. *Pracov. Lek.* 29:47-50.

Vandervort, R., and Cromer, J. 1975. Health hazard evaluation/toxicity determination report, Peabody Galion Corp. (Unisteel Body Division), Lima, Ohio. Cincinnati, OH: National Institute for Occupational Safety and Health.

Vandervort, R., and Polakoff, P.L. 1973. Health hazard evaluation/toxicity determination report 72-84-31, Dunham-Bush, Incorporated. Cincinnati, OH: National Institute for Occupational Safety and Health.

Viali, A.; Cano, J.P.; Gouezo, F.; Sauve, J.M.; Bourbon, P.; and Mallet, B. 1978. Approache methodologique du dosage du plomb, du cadmium et du chrome dans les poussieres atmospheriques par spectrophotometric d'absorption atomique sans flamme. *Pollution Atmospherique* 78:115-119.

Vitums, V.C.; Niles, N.R.; Borman, J.O.; and Lowry, R.D. 1977. Fibrose pulmonaire et silice amorphe; a propos de trois cas etudies aux Etas-Unis chez des ouvriers metallurgistes. *Cahiers. de Notes Documentaires* 89:465-469.

Vorontsova, E.I., and Pryadilova, N.V. 1976. Hygienic evaluation of copper welding. *Gig. Tr. Proj. Zabol.* 7:49-51.

Weir, F.W. 1979. Health hazard from occupational exposure to metallic copper and silver dust. *Am. Ind. Hyg. Assoc. J.* 40:245-247.

Weymouth, L.A., and Loeb, L.A. 1978. Mutagenesis during *in vitro* DNA synthesis. *Proc. Natl. Acad. Sci. USA* 75(4):1924-1928.

Wilhelmsen, L.; Oxhuj, H.; Bake, B.; Rindy, A.; and Selin, E. Investigations on effects of electric arc welding in a Swedish shipyard: A preliminary report. *Bull. Inst. Marit. Trop. Med. Gdynia* pp. 45-52.

Wisniewskia-Hejka, Z.; Cempel, M.; and Nyka, W.M. 1978. Neurological examinations of workers with chronic exposure to manganese dioxide in production of piles and batteries. *Neur. Neurochir. Pol.* 12:435-441.

Zorn, H.; Stiefel, T.; and Diem, H. 1977. Die Bedeutung des Berylliums und seiner Verbindungen fur den Arbeitsmediziner p. 2: Mitteilung. *Zbl. Arbeitsmed.* 4:83-88.